Whys and Ways of Science

Science moves, but slowly slowly,
creeping on from point to point.
Alfred, Lord Tennyson

Science is organized knowledge.
Herbert Spencer

Science without religion is lame,
religion without science is blind.
Albert Einstein

Whys and Ways of Science

Introducing Philosophical and Sociological Theories of Science

Peter J. Riggs

MELBOURNE UNIVERSITY PRESS

1992

First published 1992
Printed in Australia by
Australian Print Group for
Melbourne University Press, Carlton; Victoria 3053
U.S.A. and Canada: International Specialized Book Services, Inc.,
5602 N.E. Hassalo Street, Portland, Oregon 97213–3640
United Kingdom and Europe: University College London Press,
Gower Street, London WC1E 6BT, UK

National Library of Australia Cataloguing-in-Publication entry

Riggs, Peter J.
Whys and ways of science: introducing philosophical and sociological theories of science.
Bibliography.
Includes index.
ISBN 0 522 84471 5.
1. Science—Philosophy—History. 2. Science—Social aspects—History. I. Title.
501

To
Tracey,
Rosalind,
and Eliza-anna

Contents

Illustrations

Preface

The primary objective of this book is to present philosophical and sociological theories of science in a manner which will (it is hoped) be intelligible, informative and interesting to the reader. Many of the texts about the nature of science are either too thin on detail or too narrow in scope. Most texts do not include the more recent advances in philosophy and sociology of science. This book was written with the aim of rectifying these shortcomings.

Much of this book was written whilst I was a Visiting Research Fellow in the Department of History and Philosophy of Science, University of Melbourne. I particularly wish to express my gratitude to a number of friends and colleagues: Homer Le Grand, who was a constant source of encouragement in addition to commenting on most of the manuscript; Howard Sankey, Keith Hutchison, John Clendinnen and Philip Pettit for helpful discussions on particular issues; and also to Ms Renae Stoneham and Ms Valentina Moisiadis, both of whom patiently typed sections of the manuscript. Ms Eleanor Seymour cheerfully provided an X-ray of herself for illustrative purposes. The Australian National University provided the Bubble Chamber photograph. The staff at Melbourne University Press were always helpful, especially Ms Susan Keogh. The Graphic Design and Photography departments of both the University of Melbourne and La Trobe University are also to be thanked for producing the figures and illustrations.

Some special friends (they know who they are) deserve much gratitude for their unfailing interest and moral support.

Melbourne, 1992 — Peter J. Riggs

1
Science and Its Philosophy

In Greek mythology, it is told that the god Prometheus stole fire from heaven to give to humanity and that Zeus, the most powerful of all the gods, punished him for this action. We hold very different beliefs today, but are we really very much better off in regard to knowledge about the world than were the ancient Greeks who wrote of Prometheus' crime? Views about what science is or what exactly constitutes a body of scientific knowledge have undergone great changes over the last two millennia. Once upon a time, for instance, astrology and alchemy were both considered and practised as sciences. They would scarcely be perceived so in modern society.

It is easy to ask the question: what is 'Philosophy of Science'? It is not nearly as simple to answer it. One might say that philosophy of science enquires into: the structure of scientific theories; the methods/procedures employed in scientific research; the manner of the choice made between different scientific theories; and the metaphysical status of scientific theories. There exists much controversy on each of these issues in philosophical circles. Indeed, the philosophical debate about science can be characterized as being conducted in different 'schools' of philosophy and sociology of science. These schools are distinguished from one another by advocating a particular theory of what constitutes science or of the way in which science is supposed to work. Let the reader beware! There is little agreement among the different schools. Instead one finds only a variety of opinions, arguments and theories, some good, some not so good. Regardless of this, there is much grist for the reader's philosophical mill and a whole world of 'intellectual fodder' for the reader to digest, to criticize and to enjoy. This is the *stuff* of which philosophy of science is made.

Traditional questions asked by philosophers in their studies of science have included:

- what is (are) the aim(s) of scientific research?
- is scientific research conducted in a rational manner?
- is there a single, correct method for the conduct of science?
- is there progress in science? If so, what does it constitute?
- is there consensus/agreement in science? If so, to what degree?
- can science be consistently distinguished from other human endeavours?
- is scientific knowledge different from or superior to other forms of knowledge?
- is science successful? If so, why?
- are scientific theories true?

We shall (at least) touch on each of these questions and cover some of the answers offered by the different schools of philosophy and sociology of science.

A certain philosopher once described philosophy of science as a subject with a great past.[1] Let's begin by briefly looking at just a few of its highlights. The Greek philosopher Aristotle (384–322 B.C.) proposed a method for the conduct of science and this view of scientific method was held in high regard for more than one and a half thousand years! Francis Bacon (1561–1626), one-time Lord Chancellor of England, was to overthrow the established authority of Aristotle's doctrine on science. Bacon's writings on the philosophy of science gained him the (unofficial) title of 'Father of The Scientific Method'. (We shall examine Bacon's conception of scientific method later in this chapter.)

The twentieth century, by contrast, has seen an intellectual 'explosion' in the fields of philosophy and sociology of science. Early in the century the view of science called 'Logical Empiricism' (or 'Logical Positivism') was strongly held. In Europe, this was identified with a philosophical collective known as the Vienna Circle. Members of this group included Kurt Gödel (1906–78) and Rudolf Carnap (1891–1970). Logical empiricism was introduced into Britain by the English philosopher A. J. Ayer (1910–89) in 1936. The central thesis of logical empiricism was the 'verifiability principle'. This principle stated that propositions about the world are meaningful only if they can be empirically verified. In other words, it is only meaningful to talk about those features of the universe that can be ultimately observed by the human senses.[2]

The next most influential wave of thought on science was due to the work of the Austrian philosopher Sir Karl Popper, whose ideas about the nature of scientific research became fairly dominant in Britain. Popper argued against many of the tenets of logical empiricism, including the verifiability principle. He initially proposed that science works by a process of falsification in which a scientific theory is refuted (or falsified) by observations that are contrary to the predictions of the theory. Popper distinguished scientific theories from other theories on the basis of whether or not a theory could possibly be falsified.[3] Popper later refined his original view of science by combining a more sophisticated version of 'Falsificationism' with a blend of intuitive hypothesizing. The result was called the 'Method of Conjecture and Refutation'.

The biggest upset in philosophy of science can be laid at the feet of the American socio-historian, Thomas S. Kuhn. He turned the whole field of philosophy of science 'topsy-turvy' when he described in 1962 how a scientific theory may be abandoned by most scientists and for reasons that are not necessarily fully rational. He labelled such events as 'scientific revolutions'. Although many of Kuhn's key ideas can be traced to earlier writers, the collective presentation of these 'revelations' about science had a devastating effect. Kuhn has been followed by many others. In philosophical circles, the theories of Imre Lakatos and Larry Laudan are particularly relevant. In the more broadly based sphere of social accounts of science, the ideas of Robert Merton, Paul Feyerabend, Bruno Latour and those associated with the Edinburgh School of Science Studies have generated much interest. In the chapters which follow we shall pass a critical eye over the principal ideas about science expressed by these twentieth-century philosophers and sociologists.

The rest of this chapter will address a number of short, preliminary topics. Firstly, we shall briefly consider the nature of logical inference, together with Bacon's historically important conception of scientific method. This will be followed by an introduction to a more modern theory which was widely accepted as the correct characterization of science—the 'Standard View of Science'. In the last sections of this chapter, scientific methodology will be defined and two crucial theses in the philosophy of science will be introduced: the thesis of the 'Underdetermination of Theory by Data' and the thesis of the 'Theory-ladenness of Observation'.

DEDUCTION, INDUCTION AND BACON'S SCIENTIFIC METHOD

In formal logic there are two distinct types of arguments: deductive and inductive. Deduction is the process that is usually associated with logical argument (and with many fictional detectives). A deductive argument works from given premises to a conclusion via valid rules of inference. The conclusion of a valid deductive argument follows from its premises. It is correct to say that the conclusion of a valid deductive argument is, in a sense, contained within the premises of the argument. Moreover, if the premises of a valid deductive argument are true then the conclusion that follows must also be true. Below is an example of a simple deductive argument.

> Premise 1: If a pig has wings then it can fly.
> Premise 2: Tom's pig has wings.
> Conclusion: Tom's pig can fly.

It can be seen from this example that the stated conclusion follows from the two premises (quite independently of whether the premises are true or not). If, in addition, the premises are true then the conclusion *must* also be true. The validity of the schema (that is, the form) of a deductive argument does not depend on the truth or falsity of its premises (that is, their truth-status). A deductive argument may be formally valid but its conclusion false if one or more of its premises are false.

In contrast, the conclusion of an inductive argument does not follow necessarily from its premises, even if those premises are true. Consider now an example of a simple inductive argument.

Premise 1: Eugene's dog has fleas.
Premise 2: Helen's dog has fleas.
Premise 3: Nick's dog has fleas.
Premise 4: Monica's dog has fleas.
Premise 5: Keith's dog has fleas.
Premise 6: Renae's dog has fleas.
Premise 7: John's dog has fleas.
Premise 8: Jane's dog has fleas.
Premise 9: Rod's dog has fleas.
Premise 10: Maria's dog has fleas.
Conclusion: All dogs have fleas.

This conclusion does not follow necessarily from the premises because the content of the argument's conclusion is greater than that of its premises. In other words, it might just happen that all dogs do have fleas but this is not guaranteed by the truth of premises 1–10. The only conclusion which follows necessarily from the truth of these premises is that the individual dogs mentioned have fleas. This latter conclusion is, however, a deductive conclusion since it provides no more information than the ten premises do. The conclusion that 'all dogs have fleas' involves taking a step beyond the given premises (beyond the information provided). It can be seen, then, that the conclusion of an inductive argument is a generalization made from the specifics of its premises.[4]

When used as a method for scientific research, induction becomes part of a procedure that begins with the collection of large amounts of data—facts about the world. A sufficiently large sample of facts collected under different and varied conditions allows a researcher to identify trends (or patterns) in the data. These trends are assumed to be due to the regulation of the phenomena by laws of nature. Suppose, for example, that at a given time a researcher noticed that all materials observed to that date which conduct electricity were made of metal. If the researcher was satisfied that the sample of observations made was sufficiently large and varied, then an inductive generalization which could be made from these observations is that 'all conductors of electricity are metals'. Examples of this sort tend to conjure up questions about the possibility of error. One might well ask: what guarantees that scientific induction leads to true results and to correct scientific laws? If we grant that induction is a suitable scientific method, what then constitutes a sufficient amount of data or sufficient variation of conditions for inductive purposes? If a law is inferred from the inductive process but later new data is discovered that shows that the law does not have universal validity, does this not undermine induction as a proper method for scientific research? (In the above example such a contrary fact would be that the non-metallic element carbon conducts electricity.) Such questions as these are collectively referred to by philosophers as the 'Problem of Induction' and are central issues in epistemology. ('Epistemology' is the philosophical discipline of the study of knowledge). The distinguished British philosopher Bertrand Russell gives a lucid discussion of scientific induction in his *Problems of Philosophy*:

> The mere fact that something has happened a certain number of times causes animals and men to expect that it will happen again . . . We have therefore to distinguish the fact that past uniformities *cause* expectations as to the future, from the question whether there is any reasonable ground for giving weight to such expectations after the question of their validity has been raised . . . The belief in the uniformity of nature is the belief that everything that has happened or will happen is an instance of some general law to which there are *no* exceptions . . . The business of science is to find uniformities . . . In this search science has been remarkably successful, and it may be conceded that such uniformities have held hitherto.[5]

Francis Bacon was aware of the problems inherent in making inductions about the world. Bacon was highly critical of Aristotle's scientific method (which was well accepted up to Bacon's time) for this very reason. Bacon detailed his ideas on scientific method in *Novum Organum*, first published in 1620. (The original *Organum* was a collection of Aristotle's works compiled in the Middle Ages. The presence of the word 'novum' meaning 'new' in the title of Bacon's book implied that Bacon saw his work on scientific method as superseding Aristotle's.)[6] Aristotle had said that scientific induction proceeds by a process of enumeration of individual cases.[7] Such enumeration is just an application of simple induction, as in the example given earlier. In Bacon's opinion, Aristotle's method was flawed to the extent that it allowed generalizations to be made too quickly and to be made from an insufficient data base.[8] Bacon thought that nature exhibited simple structures that could be discovered by proper scientific investigation. He called these structures 'Forms'.[9] He argued that the correct method to be employed by researchers in order to gain true knowledge must involve the process of induction, but only through careful application so as to avoid errors. The method proposed by Bacon takes into account the possibility of counter-examples and reaches its conclusion by a series of sound (firm) inductions, not just one simple induction. In Bacon's words:

> There are and can be only two ways of searching into and discovering truth. The one flies from the senses and particulars to the most general axioms . . . The other derives axioms from the senses and particulars, rising by a gradual and unbroken ascent, so that it arrives at the most general axioms last of all. This is the true way . . . [10]

Bacon acknowledged that there exist prejudices which plague all investigators and which, he argued, must be rendered harmless. He claimed that these presuppositions, which he called 'Idols', fell into four classes: 'the Idols of the Tribe', the 'Cave', the 'Market-Place' and the 'Theatre'.[11] The idols of the tribe arise from human tendencies to suppose more regularities and order in nature than actually exist; to make generalizations too quickly; and attach undue weight to positive findings—to place too much significance on the confirming instances of a phenomenon.[12] The idols of the cave are attitudes to experiences perceived by the individual due to his or her particular constitution (physiological or mental), to education and habit, or just by accident. In other words, the idols of the cave refer to subjective interpretations we make on the experiences we perceive. The idols of the market-place are claimed by Bacon to be the most troublesome.[13] These idols are distortions and errors that appear when words in common usage are employed to explain phenomena, the concept of which requires more than the words can convey. Lastly, the idols of the theatre consist in the views and methods of established authorities or accepted philosophies. The difficulty here is for the researcher to avoid accepting such views and methods merely because they are accepted by one's peers or because they come from a recognized authority. Bacon points to Aristotle's doctrines as one example of an idol of the theatre.[14]

Bacon's scientific method was intended as a means to avoid the pitfalls of these 'idols'. The method begins with an active search and collection of data. The data assembled is then used in the construction of three lists (or tables)—Bacon's 'Tables of Presentation'. He called these the 'Table of Affirmation' (or 'Presence'), the 'Table of Negation' (or 'Absence') and the 'Table of Comparison'.[15] In the table of affirmation were listed the known instances of a single phenomenon; for example, all discovered instances of entities or objects that give out heat. The table of negation listed the instances recorded in the first table but in which the phenomenon of interest was absent; for example, an object listed in the table of affirmation but that (on at least one occasion) was cold. The table of comparison listed variations and comparisons from the other two.

The basic idea of constructing these tables was to establish true correlations for the aspects of a phenomenon under study and not just correlations that occur merely by accident. True correlations offered a 'signpost' since the 'Form' under investigation was always assumed to be present when the phenomenon was present (the form of heat is present when heat is given off). Bacon's method then proceeded by a series of inductions from less general correlations to ones of greater generality.[16] This is what Bacon meant in the above quotation when he claimed that the true way to knowledge was to derive axioms from particulars by gradual ascent. This constitutes the inductive part of Bacon's method. In a historical biography of Bacon's life and work, Paolo Rossi writes:

> [Bacon] devised the tables or instruments of classification to organise reality and thus enable the memory to assist intellectual operations ... Bacon's definition of his method [was] as a 'thread' guiding mankind through the 'chaotic forest' and 'complex labyrinth' of nature.[17]

Deduction also has a role to play in Bacon's method, although only a highly limited one. After arriving at general principles, these might then be confirmed by way of their deductive consequences, that is, from a general law one can deductively infer some specific event will occur. In this manner, Bacon thought that scientific research works from observations to general laws that in turn are used to deduce the existence of observable phenomena.[18]

Bacon's method was a milestone in the philosophy of science for it was an attempt to describe science as a systematic enterprise based upon logical principles and unhampered by past prejudices. Why then is Bacon's method not accepted today? Bacon had greatly improved upon Aristotle's method but his account of science was still problematic. The 'Idols', of which scientists needed to purge themselves, proved impossible to remove totally. Bacon had not solved the problem of induction either. (Philosophers have continued working on this very problem right up to the present day.)[19] In the use of the tables of presentation, Bacon himself found that they did not provide much progress towards making discoveries because they were neither complete enough nor were the terms used in them to describe observable entities sufficiently well defined.[20] He also neglected to incorporate the use of mathematics as an instrument for the theoretical development of science. Perhaps the most important reason of all for not embracing Bacon's method is that it fails to describe accurately the workings of scientists even in Bacon's own time, such as Copernicus or Kepler.[21]

THE STANDARD VIEW OF SCIENCE

The 'Standard View of Science' (sometimes referred to as 'Inductivism')[22] is a general account of science and scientific method. It has been the view of science commonly held by the public at large and is often mistakenly identified with the inductive part of Bacon's method. The standard view sets out the aim of science, its appropriate methods and the epistemic status of scientific knowledge. It has been the persistence of the standard view that is mostly responsible for the continuing belief that there exists a correct method for the conduct of scientific research—'The Scientific Method'—by which the laws of nature can and are discovered. There are several versions of the standard view but the overall picture of science portrayed is common throughout these versions. A suitable description of this overall picture is provided by Scheffler in his *Science and Subjectivity*:

> [it is] a common philosophy of science with independent roots . . . which has attained the status of a standard view, largely shared by reflective scientists, technical philosophers, and the educated public alike, and laying great emphasis upon the objective features of scientific thought.[23]

The ultimate aim of science in the standard view is to discover truths about the external world. Scientific research is conducted by impartial investigators who conduct their scientific activities in a manner dictated by logic and by empirical facts. Moreover, the results of scientists' work are laid open to public scrutiny, by which their results may be examined and criticized to ensure objectivity and correctness.[24] According to the standard view, two types of scientific laws are developed: 'Observational' and 'Theoretical'. Observational laws are discovered by the process of inductive generalization from data accessible to the human senses. This data results from observation by open-minded researchers and such data must also pass the very high standards of scientific scrutiny. Observational laws therefore reflect true regularities in nature since subjective factors are removed by the method of discovery and scrutiny.

Theoretical laws, on the other hand, refer to non-observable entities and processes. They do not have an empirical basis as do observational laws. As a consequence, theoretical laws may have to be changed from time to time as new evidence comes to light. This is not seen as a difficulty in the standard view. A new theoretical law will account for all the empirical facts for which its predecessor accounted and more. Scientific research, therefore, is cumulative as regards its empirical content. Scheffler summarizes these points as follows:

> When one hypothesis is superseded by another, the genuine facts it had purported to account for are not inevitably lost; they are typically

passed on to its successor, which conserves them as it reaches out to embrace additional facts. Thus it is that science can be cumulative at the observational or experimental level, despite its lack of cumulativeness at the theoretical level.[25]

This perspective of science is a rather naive one and has serious problems. In the standard view scientists are depicted as unbiased, impartial investigators who do not have any preconceived notions about the laws of nature for which they are searching. If this were the case then it would be inappropriate for scientists to receive the sort of professional training that their particular disciplines provide. Such specialized training introduces all sorts of bias (biases which Bacon labelled 'Idols of the Cave' and 'Idols of the Theatre'). If the standard view were correct on this point, a scientist's education should more closely resemble the training appropriate for general investigative work, namely the collection of facts and clues untainted by any assumptions and the piecing together of such facts into a coherent picture. Scientific education is not of this general form.

The 'Problem of Induction' remains unanswered by the standard view. What justifies the use of an inductive method? How much data is required for sound inductions? What measures can guard against the possibility of counter-instances? These questions are not addressed by the standard view of science. The standard view also does not take into account aspects of actual scientific practice evident from the history of science. In particular, the history of science shows that researchers do not always produce observational laws by inductive inferences from sets of facts. In addition, it is regularly the case that not all scientists will accept all available observational data as correct. These (and other) aspects of actual scientific practice contradict central tenets of the standard view.

UNDERDETERMINATION OF THEORY BY DATA AND SCIENTIFIC METHODOLOGY

It is obvious that scientists do need to have some method(s) of going about their research. The knowledge of the procedures appropriate for a given scientific discipline would normally be gained as part of the professional training supplied to scientists by their particular field. The literal meaning of the word 'methodology' is the study of method. (Note the distinction between 'method' and 'methodology'.) In philosophy of science, the term 'methodology' is usually taken to include the criteria by which scientists choose a particular theory from several alternatives.

What use do scientists have for such methodologies? Why cannot theory choice be determined by empirical considerations alone? In other words, why not choose the theory which has a greater degree of agreement with the experimental evidence? Surely this would be the correct course of action! It might well be —if there is only one theory in a collection of theories that is best supported by available evidence. In everyday language such a situation would be described as having only one theory which best 'fits the facts'! It is, however, not usually the case for a single theory to be in such a position, assuming that all researchers in the relevant discipline accept all the available data as correct. In science, it is commoner for there to be several theories which are equally well supported by the results of experiment (or at least by those results that are accepted by most researchers). The empirical data are of little or no assistance in singling out one theory for further research. In the terminology of philosophy of science we would say that in the case of rival and empirically equivalent theories, the experimental evidence does not uniquely determine the scientist's choice between the theories concerned. In other words, theory choice is underdetermined by a finite set of data.

Granting that experiments do not pick out one theory as the best creates the problem of how to choose between theories. It is at this point that scientific methodology enters. In this relevant sense, a methodology is a rule or criterion (or a set of rules or criteria) which tells a researcher how to choose one theory out of a collection of competing theories. Let's be perfectly clear on this point: if, of a number of theories in a collection, each independently provides results in agreement with the available experimental evidence, then a scientist must rely on some non-empirical means (that is, methodological rules acceptable to his or her discipline) in order to choose between competing theories.

One of the most frequently used methodologies in science is

that of simplicity. When there are several empirically equivalent theories available to explain some phenomenon then the researcher will choose the theory which is the simplest. The problem now is how we should characterize 'simplicity', as this can take on different interpretations. One group of researchers might take 'simplicity' to mean that a theory gives descriptions in terms of less physical variables than its rivals. Another group might take it to mean that a theory's mathematical expression is less complicated than that of its rivals. A methodology of simplicity is very attractive but such appeal does not entail that its use is appropriate. In effect, by using a methodology of simplicity, the scientist is imposing a particular view on the structure of laws of nature. Yet there is nothing to suggest that the laws of nature must correspond to any researcher's idea of what constitutes theoretical simplicity. As attractive as the notion of simplicity may seem, nature need not be so constrained.[26]

The underdetermination of theory by data also has implications for situations where there are experimental results but no theories have yet been proposed to explain these results. Given any sample of data, a large number of theories (possibly infinite) of varying degrees of complexity can be inferred. Consider the following example. The Simple Pendulum consists of a light, inextensible cord (or rod) one end of which is attached to a supporting body in such a manner that it can pivot forwards and backwards. The other is free to swing and has a mass (the bob) attached to it. The bob is only allowed to swing through a small angle. This arrangement is shown in Figure 1. Suppose that a small number of experiments are done with different lengths (L) of the pendulum and the period (T—the time for one complete swing to and fro) is measured on each occasion. These results are displayed in Table 1.

Given the data in Table 1, we can attempt to find a relationship between L and T by plotting this data on a graph and seeing if some line or curve will fit (pass through) all the points on the graph. A glance at Figure 2 shows four possible curves which fit. Indeed, an infinite number of curves could (theoretically) be made to pass through these points. All curves, or relationships, which fit a given sample of data are, logically speaking, on an

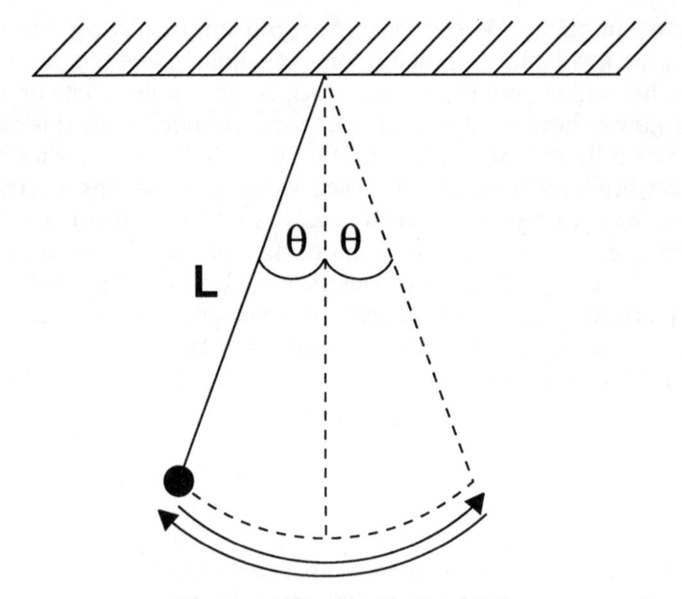

Figure 1 The Simple Pendulum

equal footing. More generally, this conclusion can be stated as follows. On the sole basis of a finite set of data, each and every possible theory consistent with the data has an equal claim to explaining it. What should be done now? As outlined above, some methodology must be invoked to single out a particular theory. In examples of this sort, the most common methodological rule employed is to choose the 'simplest' curve (the smoothest). The logician Willard Quine explains why this is so:

> If two theories conform equally to past observations, the simpler of the two is seen as standing the better chance of confirmation in future observations. Such is the maxim of the simplicity of nature. It seems to be implicitly assumed in every extrapolation and interpolation, every drawing of a smooth curve through plotted points.[27]

An alternative response to the underdetermination by a given sample of data is to enlarge the sample. Additional data points should help to decide the choice of curve. One might then return to the pendulum and conduct further tests. Each time the

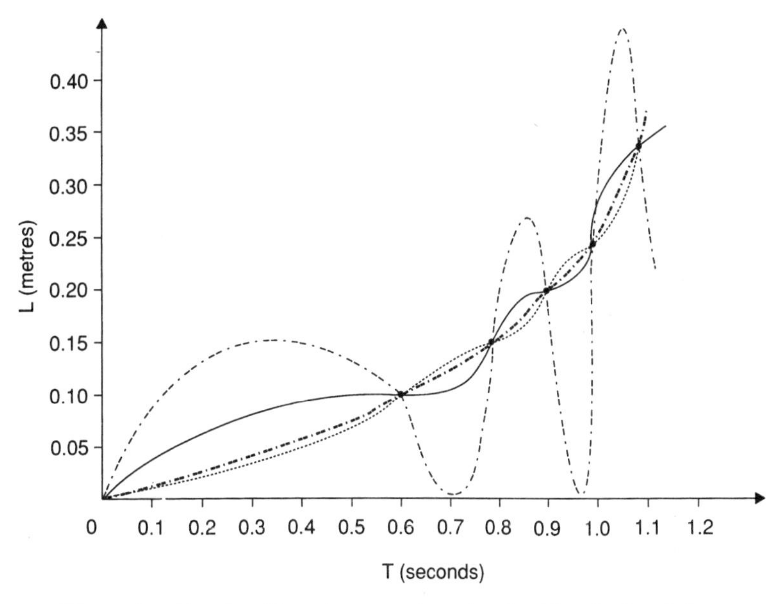

Figure 2 Graph of length verus time for the Simple Pendulum

L (metres)	0.10	0.15	0.20	0.25	0.35
T (seconds)	0.6	0.8	0.9	1.0	1.1

Table 1 Data values for Figure 2

data pool is added to, some (but not necessarily all) of the curves which fitted previously are eliminated. However, more new curves that fit are found as well. This is regardless of how many lengths of the pendulum are tried. The collection of extra data does not seem to be of much assistance in deciding which curve represents the relationship between T and L. This is a different form of underdetermination of theory by data. In this case, theory choice is not uniquely determined by any set of data.

In the standard view of science, the method for discovering observational laws is scientific induction. It is assumed in the standard view that only one theory is generated at any one time by this method and therefore there is no need for rules to decide between rival theories. However, any number of theories can be inductively inferred from a finite sample of data. In order to overcome this difficulty, it is implicit in the inductive process of the standard view that the 'simplest' theory possible will result from a 'correct' induction. In the above pendulum case, the smoothest curve (with the smallest number of terms in the equation that describes it) is taken as being the simplest.[28] Thus even in the standard view of science there is a need for a scientific methodology.

THEORY-LADENNESS OF OBSERVATION

The thesis that there cannot be a clear-cut distinction between fact and theory is called the theory-ladenness of observation. This thesis asserts that there are always some preconceptions that 'colour' our outlook of the world around us. These preconceptions are due to many factors: individual physiology, personal psychology, and the attitudes of one's society, for example. It is not even necessarily the case that we all 'see' the same thing when looking at the same object. One example familiar to us all is our own reflection in the mirror. When we look into a mirror we do not 'see' ourselves as others see us. The image of one's body that appears on the retina of one's eye may be identical with the image on the retina of another person's eye, yet both people do not perceive exactly the same thing. One tends to 'see' one's self in a better light than others do. We look at ourselves and we perceive a picture of the self which may be happier, healthier, thinner, prettier, or stronger than we actually are. (This effect is reversed in some people suffering from anorexia who, despite being quite thin, 'see' themselves as overweight.) Our minds do have a tendency to let us 'see' more of what we would like to see than is actually there.

The proposition that the human brain alters the optical image received on the eye's retina is not controversial. The brain can readily make changes or compensations, as has been conclusively demonstrated by a number of well-known and documented experiments involving inverting lenses and mirrors.[29] Perhaps the most striking of these compensations is the one continuously made by the brain whilst we are seeing. The image on the eye's retina is upside down with respect to the external object being looked at and the brain alters what we perceive so that the object appears to be rightway up!

Every day we find theory-ladenness in the form of low-level theoretical assumptions. These assumptions are implicit in even the most basic 'observational' statements. They contain a minimal amount of theoretical content and usually do not require in depth specification: for example, water is wet. Low-level theoretical definitions may simply be given ostensively. (An ostensive definition is made by indicating or otherwise pointing out the thing that one wishes to define.) A graphic example of low-level theory-ladenness is provided by the anthropologist Colin Turnbull, in his book *The Forest People*. Turnbull had made an extensive study of the pygmy inhabitants of the Ituri Forest in the Congo region of Africa. These people, who are born and raised in an environment where visibility is highly restricted, have not developed a capacity to perceive distant objects as being far away. One of the Ituri pygmies (named Kenge) accompanied Turnbull to an open area beyond the tree line. Turnbull describes what happened in the following extract from his book:

> And then he [Kenge] saw the buffalo, still grazing several miles away, far down below. He turned to me and said, 'What insects are those?'
>
> At first I hardly understood, then I realized that in the forest vision is so limited that there is no great need to make an automatic allowance for distance when judging size . . . When I told Kenge that the insects were buffalo, he roared with laughter and told me not to tell such stupid lies.[30]

The theoretical component of Kenge's observation included the preconception that the perceived size of an object is its true size.

In scientific theories the amount of theory-ladenness is much greater since all scientific theories depend on a large number of theoretical premises, some low-level, others of much greater complexity—high-level theoretical assumptions. Consider Plate 1.

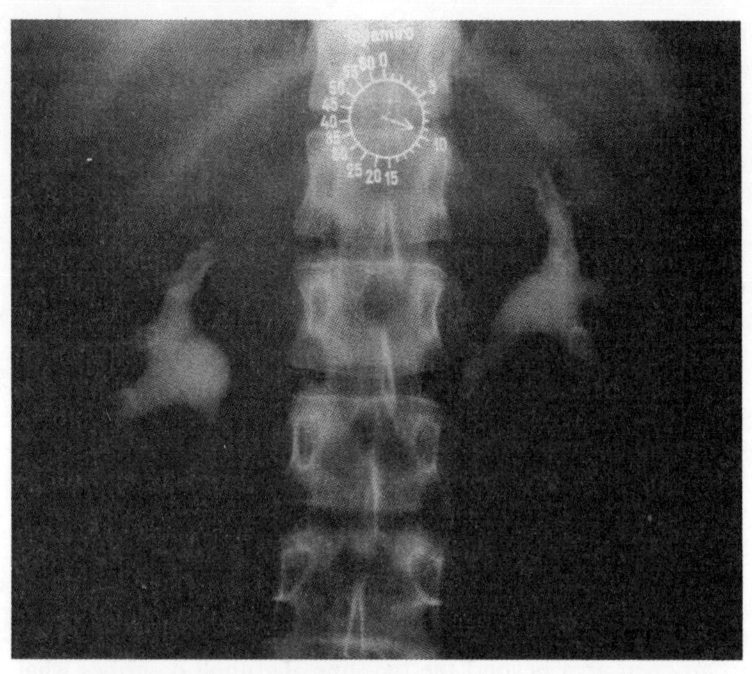

Plate 1 Blurred time-exposure photograph of two bats in a cave or an abdominal X-ray?

Is it a slightly blurred photograph? Maybe it is a time-lapse photograph of two bats flying around a tall column of rock inside a dark cave (note the superimposed clock). It could be but, of course, it isn't. Anyone experienced in viewing X-ray photographs would probably recognize Plate 1. It is an X-ray photograph of a human backbone with a kidney on either side. Anyone who is not familiar with X-ray photography would make another interpretation. Any alternative interpretation of Plate 1 depends (in part) on what is already known to the observer. What about Plate 2? It appears to be a jumble of straight and curved lines mixed up with the odd small spiral. A piece of modern art perhaps? The particle physicist would disagree, claiming that this plate shows collisions, creations and destructions of subatomic particles. The interpretation of these intersecting lines and spirals is based on a maze

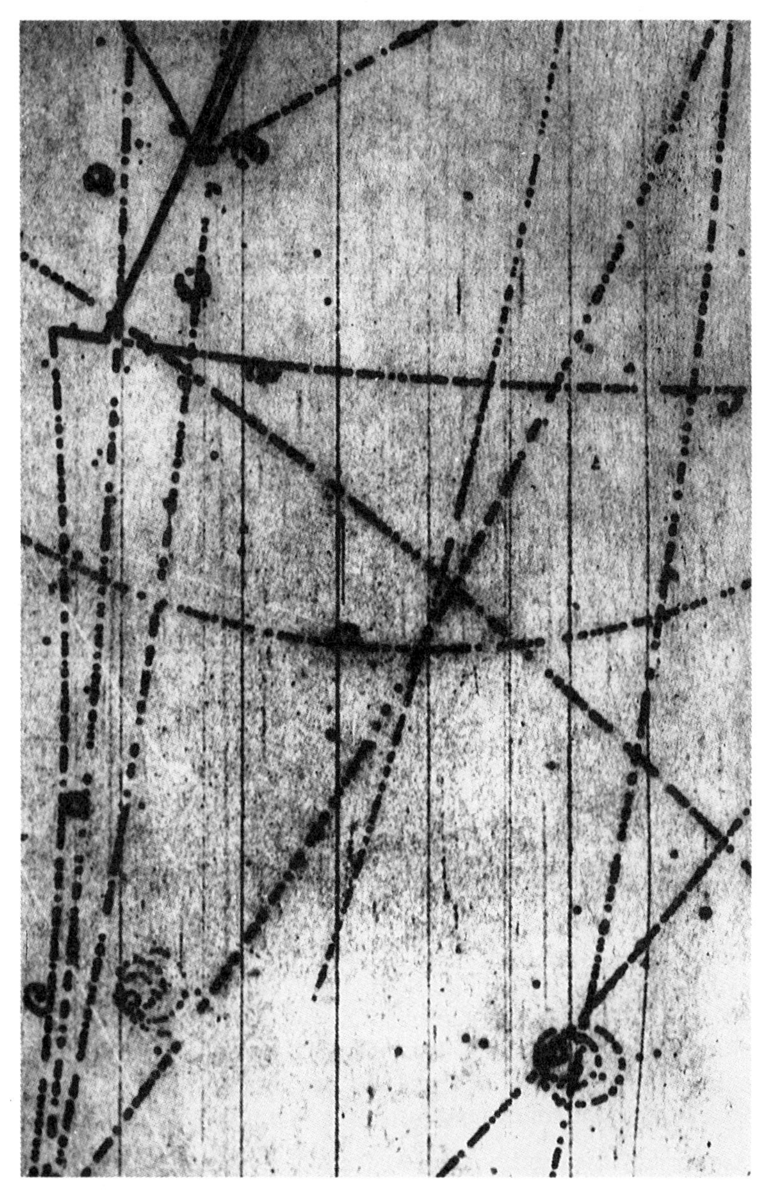

Plate 2 Abstract art or particle physics?

of high-level theoretical premises regarding the nature of the sub-atomic region, the laws of physics and the workings of the device used to record the particle tracks. In the absence of this knowledge, Plate 2 is nothing but a collection of lines and spirals.

There are two main versions of the thesis of theory-ladenness, which shall be denoted as 'Hard' and 'Soft'. In the soft version, all observers with normal vision will see the same thing when looking at the same object. They will, however, make different interpretations of these retinal images. The hard version states that it is not the case that different observers make different interpretations of what they see; instead these observers see *different things* when they look at the same object. This is sometimes expressed by saying that different observers live in 'different worlds'.

It was noted earlier (p. 12) that not all researchers will usually accept all available experimental data. Theory-ladenness is one reason that accounts for this. In the 1770s, for example, the 'Phlogiston Theory' was challenged by Antoine Lavoisier's 'Oxygen Theory' of combustion. When in about 1775 it became possible for chemists to prepare an 'air' (oxygen) that supported burning much better than 'normal air', this apparent matter of fact turned out to be different 'facts' to different chemists. To the phlogistic chemist Joseph Priestley, this gas was de-phlogisticated air.[31] The same gas separated by the same process was, to Lavoisier, the 'Principle of Acidity' (which, he claimed, in combination with the substance of heat—caloric—gave oxygen). This is just one example from the history of science of theory-ladenness of observation.[32]

Theory-ladenness of observation raises the question of whether there can be a theory-neutral (or independent) observation language for science. An observation language is one wherein the results of experiments are presented, recorded, communicated or otherwise utilized. Observation statements appear only to be expressible in terms of some existing theory. J. J. C. Smart offers the following illustrative example:

> observation statements are *not* statements about pointer readings. A scientifically untutored peasant could certainly make a report that a

> black needle-like thing pointed to the figure "35" on a round clock-like thing. He could *not* report that the current through a milliammeter was thirty-five milliamperes, since he would not have the concept of an electric current, and still less would he have the concept of an ampere.[33]

How can the reading on a scientific instrument have any meaning in the absence of a scientific theory that explains it? If there is no clear distinction between facts and theory, how can there be any theory-neutral language? No attempt to produce a theory-neutral language has been successful and although this does not show it to be an impossibility, nevertheless the prospect of a theory-neutral language must be considered highly improbable.[34]

2
Kuhn's Theory of Scientific Revolutions

Thomas S. Kuhn, in his book *The Structure of Scientific Revolutions* (originally published in 1962), put forward a radical description of science which provoked lengthy and intense discussion. The impact Kuhn's theory had in the areas of the history, philosophy and especially sociology of science came as quite a surprise. (Kuhn added a postscript to the book in 1969 which appears in the second edition as a partial response to his critics. In this chapter we shall be solely concerned with his original account.) Kuhn largely intended his theory to be applicable to the natural sciences as distinct from the social sciences (such as economics). He concentrated on the empirical problems of science—those concerning the lack of match between theory predictions and experimental results. Kuhn denied that scientific research was merely a cumulative enterprise which gradually built up knowledge over time and resulted in truths about the external world. Instead, he claimed that, although scientific research is quite a mundane activity most of the time, the accepted theories and methods of science can occasionally be abandoned by scientists for alternative ones in rather dramatic episodes called 'scientific revolutions'.

When Kuhn formulated his scheme of science and scientific change there was already much written about the nature of science. Why did Kuhn come to differ in several crucial respects from many of the other accounts of science? Kuhn himself provides some insight into his motives. In the Preface of his book, Kuhn outlines the reasons which helped persuade him that earlier accounts of science were either seriously flawed or incomplete. He writes:

> I was a graduate student in theoretical physics . . . A fortunate involvement with . . . physical science for the non-scientist provided my first

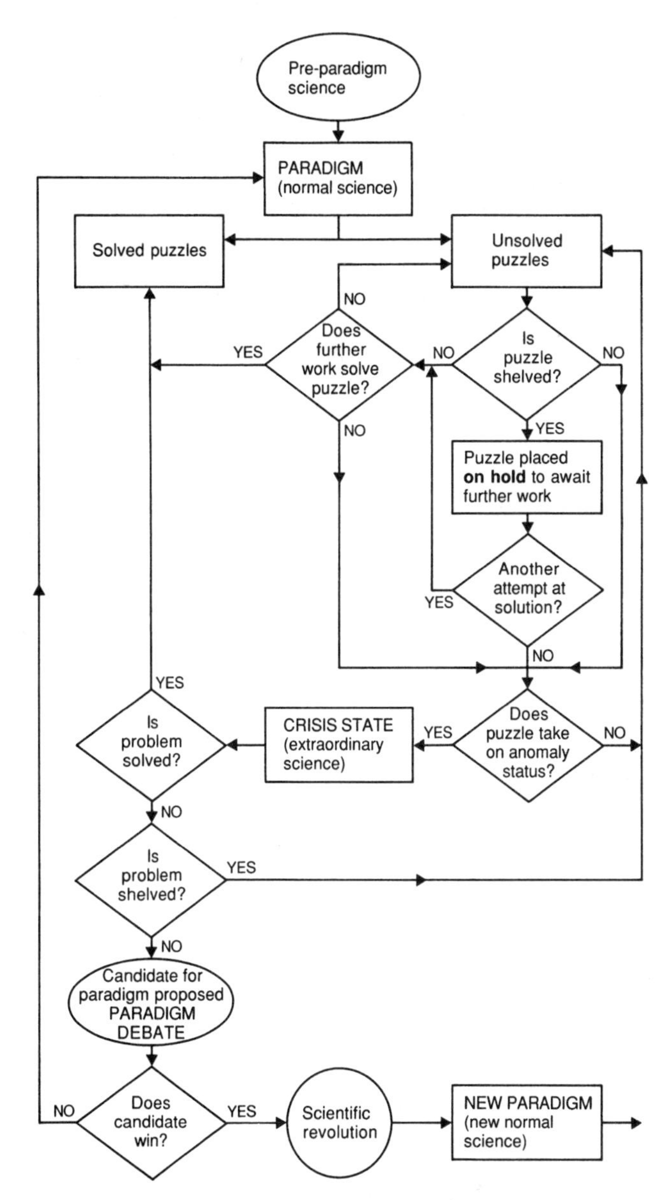

Figure 3 Kuhn's scheme of science

> exposure to the history of science ... [that] radically undermined some of my basic conceptions about the nature of science ... these notions did not at all fit the enterprise that historical study displayed.[1]

Kuhn's exposure to the history of science indicated to him that previous renderings of the nature of scientific research focused too narrowly on particular aspects of science to the exclusion of others. Kuhn's study of various historical cases led him to conclude that the scientific theories were themselves only part of the larger picture of what constituted scientific research. He portrayed this in a scheme that he believed would provide a fuller account of science than previous ones and that was strongly supported by historical evidence. Kuhn claimed that science could be displayed as a whole rather than simply in many bits and pieces, as historians tended to do. He named the holistic structure of science the 'Paradigm'. The rest of this chapter is devoted to the details of Kuhn's theory, the logical form of which is presented as a flow chart in Figure 3.

PRE-PARADIGM SCIENCE

Kuhn takes the term 'paradigm' to mean a framework (or 'umbrella') of theory, assumptions and methods under which scientists in a particular discipline conduct their research. However, Kuhn states that there is also a stage in the development of science prior to that characterized by research guided by a paradigm: 'Pre-paradigm Science'. At this stage in the development of a discipline there is no general agreement amongst researchers on the fundamentals of their field. Instead there exist several different schools of thought as to what constitutes the basis of any given field of research.[2]

Kuhn provides historical examples to demonstrate what he wishes to convey by his term 'pre-paradigm science'. He argues, for instance, that the discipline of physical optics was, before the publication of Sir Isaac Newton's famous treatise *Opticks* in 1704, in a pre-paradigm state:

> No period between remote antiquity and the end of the seventeenth century exhibited a single generally accepted view about the nature of

> light. Instead there were a number of competing schools and sub-schools, most of them espousing one variant or another of Epicurean, Aristotelian, or Platonic theory . . . Yet anyone examining a survey of physical optics before Newton may well conclude that, though the field's practitioners were scientists, the net result of their activity was something less than science. Being able to take no common body of belief for granted, each writer on physical optics felt forced to build his field anew from its foundations.[3]

A discipline in this early stage of its development is also called an 'Immature Science'.

It is not just in the history of physics that these sort of situations can be identified. We draw a second example of a pre-paradigm state from the history of biology. The publication in 1859 of Charles Darwin's *Origin of Species* greatly upset existing opinions in the field. (Darwin's impact was so great that some historians and philosophers interpret it as a revolution in science.) There were, just as in the optics case, a number of competing schools in existence such as the traditional creationist view and some alternative evolutionary themes. Heated debates ensued after publication of *Origin*, but Darwin's theory of evolution by natural selection gradually won over most scientists and journals in the field.[4]

Science is noted for a high degree of consensus amongst its practitioners. How is it then that a discipline which begins in an atmosphere of disagreement over fundamentals develops into a coherent body of knowledge? Kuhn says that one of the competing schools will, sooner or later, attract more practitioners than its rivals. If this continues to occur the rival schools can only shrink in size and eventually cease to have a noticeable effect on research in the field. What could bring this about? One way would be for the theoretical assumptions of a school to appear in some way superior to those of the competing schools.[5]

Kuhn argues that the history of science shows one school of thought always survives while the others perish. He cites the case of electrical research in the eighteenth century. Led by the investigations of Benjamin Franklin, his school (which assumed electricity to be a single fluid) was to gain the upper hand at the expense of others.[6] Franklin's arguments not only seemed better than those of his rivals but his school could also partially explain more phenomena—for example, how the Leyden Jar (a device for storing electricity) worked.[7] Franklin claimed that the Leyden Jar 'bottled' the electrical fluid. Franklin's achievements served to bring together most 'electricians' under a common set of guiding principles.

The victory of one of the pre-paradigm schools in a particular field of research over its competitors is accomplished by that school having more to offer (or at least being seen to have more to offer) the practitioners in the field than do the other schools. This results in defections from these other schools and acceptance by those new to the field of only one school's beliefs. As a consequence the rival schools decline in adherents and consequently will cease to function.[8]

NORMAL SCIENCE

The school that emerges triumphantly from the pre-paradigm stage (to the exclusion of all others) then conducts its research under its own paradigm. Kuhn calls this type of research 'Normal Science'. Let's attempt to give better definitions of the notions of paradigm and of normal science. Kuhn defines normal science as:

> research firmly based upon one or more past scientific achievements, achievements that some particular scientific community acknowledges for a time as supplying the foundation for its further practice.[9]

If a scientific achievement has been made and then provides the foundation for further research, Kuhn would want to say that normal science is that activity in which most scientists will be engaged in most of the time. Further, a paradigm can be said to be a 'concrete scientific achievement' to which Kuhn assigns the following two characteristics:

> [it is] sufficiently unprecedented to attract an enduring group of adherents away from competing modes of scientific activity . . . [and] sufficiently open-ended to leave all sorts of problems for the redefined group of practitioners to resolve.[10]

What might such scientific achievements be? Perhaps a highly successful hypothesis would fulfil this role, that is, a theory that makes many and varied predictions of observable phenomena and these predictions are then experimentally confirmed. Or perhaps it would be a piece of apparatus that allows for further investigation of some facet of nature. Kuhn seems to have these sort of ideas in mind, advocating that classic textbooks in science (such as Newton's *Opticks* or Darwin's *Origin*) acted for scientists as a focus of attention on various achievements. These works then served to establish legitimate methods and problems for the researchers in a scientific discipline.[11] Any text that draws attention to a scientific achievement and thereby implicitly defines problems and methods in a field does so because it (the text) is an exemplar. The term 'exemplar' means an accepted example of actual scientific practice.

Yet there is more to the concept of paradigm than just this. In addition to being 'an accepted model or pattern', a paradigm is also 'an object for further articulation'.[12] A paradigm embodies the theoretical and methodological principles of the research field that it governs. Acceptance of the relevant paradigm allows scientists to pursue their research without constantly questioning and debating the basis of that research. (A scientific discipline that has reached this stage is called a 'Mature Science'.) This is what Kuhn labels as articulation of the paradigm and the research performed by scientists under a paradigm is what he calls normal science.

It is all very well to speak generally of paradigm-guided research but this does not provide much detail. One might well ask: what *precisely* is done in normal science? Kuhn thinks that normal science is about solving special sorts of problems. Indeed he says that the aim of science is to solve problems.[13] The paradigm defines the type of problems that await solution, the methods to be employed in reaching solutions and the guidelines as to what solutions are acceptable. A problem only appears against a background provided by the paradigm; in other words, what constitutes a problem in science is dependent on the context. In the history of astronomy, for example, a most perplexing problem was to account for the apparent backwards motion of the planets when viewed over several months against the background of the 'fixed' stars. Most models of the Universe since the time of the ancient Greeks offered different solutions to this problem of planetary retrogression. In the Copernican system, retrogression ceased to be a problem because the observed motions of the planets were a natural consequence of the Earth, and the other planets, orbiting the Sun.

What are the special problems that scientists spend their professional lives attempting to solve? Kuhn labels such problems of normal science 'puzzles'. These puzzles are not just any sort of problem but ones where there are boundaries and/or rules (although they are perhaps implicit ones) that apply to the ways and means of their solution. Scientific puzzles are defined within the reigning paradigm and these puzzles are always assumed to have solutions. Thus researchers can set about looking for answers to puzzles without having to worry whether a solution exists or not. It is important to note that it is always assumed that solutions to normal scientific puzzles can be found:

> one of the things that a scientific community acquires with a paradigm is a criterion for choosing problems that, while the paradigm is taken for granted, can be assumed to have solutions . . . that its [science's] practitioners concentrate on problems that only their lack of ingenuity should keep them from solving.[14]

Kuhn identifies three different classes of scientific puzzles: the determination of significant fact; the matching of facts with theory; and the articulation of theory.[15] The determination of significant fact would be where a scientist attempts to increase the accuracy of the value of a physical constant (the speed of light in vacuum, say) by making more exacting measurements so as to increase the number of significant figures in the known value of the constant. The matching of facts with theory involves the direct comparison of experimental 'facts' with the predictions from the paradigm. Puzzles dealing with the articulation of theory come in two varieties. Firstly, there are those that require the resolution of ambiguities in the paradigm so that it becomes evident what the puzzle actually is. Secondly, there are those that involve the extension of the paradigm into new areas, for example, extending Newtonian physics into physiology in order to explain mechanically the workings of the body.

Kuhn states that in normal science there must be rules that govern its conduct and what counts as a suitable solution.[16] But what are these rules? Kuhn compares the puzzles of normal science to some standard everyday games and puzzles in order to draw out an analogy. Take, for example, a jigsaw. There is a set method for solving a jigsaw: the pieces have to fit together smoothly (without the corners of pieces being broken off) and the finished product has to be a particular picture, for example, corresponding to the one on the cover of the jigsaw box. Kuhn would deny that there is a unique method or approach in science. Therefore there cannot (in his view) be a single set of governing principles (rules) for science. Instead he would say that each paradigm defines its own set of rules and standards.

Despite this, the rules of normal science need not be in an explicit form. Kuhn thinks that shared paradigms are not only more important than any rules that might be extracted from such paradigms but also that the existence of the paradigm is prior to the advent of any shared set of rules.[17] This is not really surprising since the beginning of a paradigm has historically coincided with some concrete scientific achievement (such as the publication of a classic textbook). In this sense the paradigm is prior to the existence of any accepted set of explicit rules governing normal science.

Unequivocal rules are not essential since acceptance of a paradigm implies that the scientist possesses what has been termed 'tacit knowledge'—knowledge gained by the scientist during his or her professional training and which is not (or perhaps cannot be) fully expressed. Kuhn refers to this as the process of 'learning by finger exercise'; one learns by firstly observing and then repeating.[18] The existing models of scientific practice under the reigning paradigm are the exemplars from which tacit knowledge is secured.

The conduct of normal scientific practice results in a number of occurrences. Firstly, puzzles in a particular scientific discipline are acknowledged or recognized and then (mostly) solved. Secondly, the rate of solution of puzzles is usually such that the amount of information in a field expands markedly. Normal science, as a consequence, is a period of much accumulation of knowledge.[19] It has been these periods of normal science that Kuhn identifies as being the basis of the mistaken belief that science is merely a cumulative process (that is, an amassing of knowledge about nature year after year). The increase in solved puzzles and knowledge in a field is one form of scientific progress that is identifiable in Kuhn's theory of science. Such rapid progress as appears in normal scientific research is only possible due to the consensus of the practitioners on the fundamentals, the methods and practice of their field.

Kuhn's definition of the term 'paradigm' is vague and imprecise for it allows for a multiplicity of interpretations. One commentator on Kuhn's theory, Margaret Masterman, has provided no less than twenty-one different senses of the term 'paradigm'.[20] The lack of a very specific, explicit definition on Kuhn's part (at least in the original edition of *Structure*) is to his advantage in that a 'rubbery' formulation of the paradigm allows Kuhn the luxury of only producing fairly general arguments. Yet it is the paradigm that governs the conduct of normal science and although the rules of normal science need only be implicit, they can be made explicit during a crisis. The large degree of flexibility in definition of the paradigm would appear, prima facie, inconsistent with the existence of the (implicit or explicit) rules that govern normal science.

There is a further difficulty with Kuhn's description of the state of normal science. He claims that there is only a single, dominant paradigm in any one discipline at any one time (this is also referred to as the exclusivity of a paradigm). This claim is refuted by the history of science. For example, it is easy to identify a period in history when there were co-existing 'paradigms' of Ptolemaic, Copernican and Tychonic astronomy, all competing for the dominant position. The field of astronomy was not in a pre-paradigm state at this time since each of the groups which adhered to one of these 'paradigms' was engaged in research that corresponds to the description of Kuhnian normal science.

UNSOLVED PROBLEMS AND PUZZLES

Normal science under its paradigm could presumably continue indefinitely but historically this has not occurred. Despite the faith researchers have in their paradigm, Kuhn says that there will always be problems that arise that cannot be solved within its framework. The unsolved problem is a general category that includes unsolved scientific puzzles. (Recall that puzzles are the special type of problems for which acceptance of the paradigm provides the means of solution.) However, if there are problems that cannot be solved within a particular paradigm, then these cannot be classified as puzzles (in Kuhn's sense of the word). What happens with unsolved problems in science? In the first instance, nothing much happens at all. Kuhn argues that there is nothing particularly special about the existence of unsolved problems. Not all the puzzles of normal science are going to be so obliging that they surrender their secrets with only a small amount of effort on the part of an investigator. In order to solve a particular puzzle the attention of several researchers may be needed, further experimentation may be required, new instrumentation may have to be designed and so on. Such activities as these take time.

The existence of problems that remain unsolved for some time (by itself) presents no difficulties for normal science. It is taken for granted that, sooner or later, the problem will be solved:

> There are always some discrepancies [between theory predictions and observation]. Even the most stubborn ones usually respond at last to normal practice. Very often scientists are willing to wait, particularly if there are many problems available in other parts of the field.[21]

For example, for about sixty years after Newton's calculations on lunar motion were made (around 1690), the observed motion of the Moon still did not exactly agree with that predicted by Newtonian mechanics.[22] This puzzle was eventually to be solved in the 1750s within the Newtonian paradigm primarily by the French mathematician Alexis Clairaut.[23] This example indicates that an unsolved puzzle may be continued to be worked upon for long periods in the belief that it will be solved. The alternative strategy in normal science is just to 'shelve' the puzzle, leaving it to await a researcher sufficiently interested to tackle it again, or some advance in instrument technology to assist in finding its solution. Such shelving is seen as nothing out of the ordinary in normal science.

At this point in the discussion, let's digress slightly. Kuhn does not accept a clear-cut distinction between fact and theory; that is, he accepts that all observations are theory-laden. What counts as a fact may differ from one paradigm to another. Kuhn adopts a 'hard' version of theory-ladenness. It was said in Chapter 1 (p. 20) that chemists who adhered to phlogiston theory viewed the same gas separated by the same process as a different gas from those who adhered to oxygen theory. Kuhn argues that both sorts of chemists 'see' a different chemical reaction.[24] It is not that these two chemists had alternative names for the one gas; the two concepts involved are totally distinct and they lead to very different explanations of chemical reactions. In a sense, both Lavoisier and Priestley 'saw' a different chemical reaction. Kuhn's description for this is that scientists working within separate frameworks practise 'in different worlds'.[25]

We have seen that the expectation in normal science is that all problems will be solved at some time. There is also an expectation that normal research will produce answers which could be somewhat anticipated from the paradigm or are at least in keeping with its general theme.[26] Yet novelties (unexpected events) do occur in normal science. It is the task of the researcher to integrate the novelty with accepted theory smoothly. Kuhn initially labels a novelty in normal science as an anomaly, meaning something unexpected or out of the ordinary. He later changes the definition of the term 'anomaly'. To avoid possible confusion we shall refer to this sort of novelty as an unexpected result. (Where quotations from Kuhn are likely to confuse, the term 'unexpected result' will be inserted in brackets after the word 'anomaly'.) The assimilation of discovery by researchers is part of the articulation (through extension) of the paradigm. This process of assimilation begins with the finding of an unexpected result:

> Discovery commences with the awareness of anomaly [unexpected result], i.e., with the recognition that nature has somehow violated the

> paradigm-induced expectations that govern normal science. It then continues with a more or less extended exploration of the area of anomaly [unexpected result]. And it closes only when the paradigm theory has been adjusted so that the anomalous has become the expected.[27]

A case in point is the discovery of X-rays. In 1895 the reigning paradigm included acceptance of several types of radiation but Kuhn states that the discovery of X-rays was treated with much surprise and doubt.[28]

Historically, there have always been some problems in every paradigm that cannot be satisfactorily solved. These may mistakenly be thought to be merely puzzles. This is because there is no way of initially differentiating a puzzle which can be solved in normal scientific practice from a problem which cannot be so solved. Problems that have remained unsolved for long periods of time (perhaps several centuries) may begin to take on a new status. The longer a problem remains unsolved (especially after repeated attempts at solution) the more likely it will be that this problem comes to be viewed in a different light; that is, as something different from and far more serious than just a puzzle of normal science. (Although, as we shall see, this will not be the only factor present.) Any unsolved problem that previously was considered to be a puzzle but has resisted all attempts at solution and persisted to the point of causing crisis in a scientific field (that is, a state of deep concern and doubt amongst scientists) shall be called an anomaly (or a Kuhnian anomaly). This usage of anomaly agrees with Kuhn's re-definition of the term. The nature and effects of anomalies and crises in science are what we shall turn to next.

THE ANOMALY AND THE CRISIS STATE

In its full sense a Kuhnian anomaly is an apparent counter-instance to a theory; the existence of an anomaly shows that there is direct conflict between theory and observation. A counter-instance indicates, prima facie, that the theory as stands fails to give accurate predictions of phenomena and therefore needs either modification or perhaps replacement. An alternative response to the existence of an anomaly is that something is wrong with the experimental arrangement (for example, incorrect calibration of apparatus) and this is causing the appearance of an anomaly. It is quite usual, according to Kuhn, for researchers initially to view an anomaly not as a counter-instance but only as an unexpected result which will be explained at some later time. This is normal scientific practice and for good reason. Science would make little progress—solve few problems—if every unexpected result were to be treated as a counter-instance to a theory.

Sir Karl Popper's methodology of science was briefly mentioned in Chapter 1. Popper claims that a primary role of scientific research is to 'falsify' theories.[29] He claims that the best scientific theories are those that have more potential to be falsified by experiment than other scientific theories. (Our currently held theories have survived all such efforts at falsification to date.) Thus, in Popper's view, scientists continually make strenuous efforts to falsify theories. Kuhn thinks that if one were to accept this 'Falsificationist' view, then every unexpected result in science would be treated as a counter-instance and the theory concerned would be refuted.[30] This point, Kuhn thinks, is a major failure of Popper's philosophy of science for it would lead to ridiculous situations where scientific theories were being unnecessarily (and inefficiently) thrown away. A theory that provides good predictions in some circumstances may need modification in others, but not necessarily outright rejection. The well-known adage: 'Don't throw the baby out with the bathwater', comes to mind. (Kuhn actually takes a simplistic view of Popper's writings. Kuhn criticized *naive* falsificationism but Popper's philosophy involved a more sophisticated version of falsificationism.)[31]

If all unsolved problems in science are initially assumed to be merely puzzles and real anomalies are not immediately recognized as such, what happens to change the perception of the researchers involved? It is not just because a problem has remained unsolved for a period of time.[32] This we saw in relation to the example of the Moon's orbit. The problem of the solar neutrino flux is a more recent example. The neutrino is a subatomic particle that interacts only very weakly with matter.

Neutrinos are emitted in terrestrial nuclear decay processes and also from the fusion of hydrogen into helium in the core of the Sun and other stars. The rate of emission of neutrinos from the Sun is a measure of the internal temperature of the solar core. Since the mid-1960s, when measurements of the solar neutrino flux began, there has been a discrepancy between the recorded rate and the predicted rate of neutrino emission.[33] In an effort to gain a better match between these rates, more sensitive neutrino detectors have been constructed and painstaking refinements to the theory of solar reactions have been made. This has resulted in the gap between theory and experiment being narrowed, but not closed. This continuing problem has not resulted in crisis for the relevant scientific community; it has merely intensified efforts in the attempt to bring about a solution.[34]

When a crisis state does develop in science there are other reasons present (in addition to the existence of a problem or problems that have resisted solution) that act together to cause the crisis. Kuhn initially uses the following poorly phrased expression to describe the crisis inducing problem: 'if an anomaly [unexpected result] is to evoke crisis, it must usually be more than an anomaly'! Kuhn's description of anomalies does (fortunately) improve. He continues:

> There are always difficulties somewhere in the paradigm-nature fit; most of them are set right sooner or later . . . We therefore have to ask what it is that makes an anomaly seem worth concerted scrutiny . . . Sometimes an anomaly will clearly call into question explicit or fundamental generalizations of the paradigm . . . or . . . the development of normal science may transform an anomaly that had previously been only a vexation into a source of crisis . . . Presumably there are still other circumstances that can make an anomaly particularly pressing, and ordinarily several of these will combine.[35]

What is the form of these other circumstances? In the case of the Einsteinian Revolution (1905–20), for instance, Albert Einstein perceived that there were asymmetries and inconsistencies in the Newtonian paradigm. These, in part, led him to both his Special and General Theories of Relativity.[36] In the Copernican Revolution of the sixteenth century, there existed a need for an accurate calendar in which the Holy Days of the Catholic Church and the seasons of the year appropriately corresponded.[37] This need served to highlight the already acknowledged errors in the accepted system of that time, the Ptolemaic astronomical system.

In situations where an unsolved problem in a scientific field takes on a new and more important status due to both the realization that no one has managed to solve it and that there are additional factors (peculiar to each individual situation) that make the solving of the problem an urgent matter, the particular scientific field is said to be in a state of crisis. Crisis means that something has gone wrong with the normal problem-solving approach. When this occurs normal scientific practice gives way to methods that would not be approved of if the field were not in a crisis state. These new methods are characterized under the title of 'Extraordinary Science' (to distinguish it from normal science):

> [When] an anomaly comes to seem more than just another puzzle of normal science, the transition to crisis and to extraordinary science has begun. The anomaly itself now comes to be more generally recognized as such by the profession.[38]

Figure 3 (p. 23) shows that once an unsolved problem takes on anomaly status, then the field finds itself in a crisis state. Kuhnian anomalies act as 'catalysts' of crisis and the more of them present in a particular discipline, the quicker the onset of crisis.

Kuhn has difficulty in specifying the exact difference between an unsolved problem and a Kuhnian anomaly. If a serious unsolved problem does become an anomaly, Kuhn does not state how long a time period is needed before the problem takes on the status of a Kuhnian anomaly. While a greater number of anomalies quickens the onset of crisis, one might ask just how many anomalies are required? Suppose that there are three serious problems in a particular paradigm but no crisis has emerged; how is it then that perhaps the existence of only one more serious problem does bring on crisis? Kuhn offers no definite answers.

What are the reactions of scientists to a crisis state? First and foremost, scientists do not abandon their paradigm, that is, the anomaly is not interpreted as a refutation of the paradigm.[39] They continue to work within it but their research methods will not be as restricted as in the case of normal science. The rules governing scientific practice (whether they are known explicitly to the scientist or not) will be loosened in the attempt to solve the problem. Indeed crisis in science is regularly a time when implicit rules are made explicit in order to help clarify matters in the field. Adherence to a paradigm by researchers should not come as a surprise, since most will have large personal and professional stakes in the continuance of the paradigm. Many years (perhaps most of one's working life) spent solving problems in one paradigm and building up a reputation on the basis of such work provide very strong reasons for the maintenance of allegiance.

The scientist does not attempt to blame the crisis on the paradigm. Any researcher who did so would be seen in a similar light to 'the carpenter who blames his tools'.[40] Acceptance of the paradigm carries with it the assumption that failure to find solutions is due only to the lack of ability or ingenuity of the paradigm's adherents. Any failure to solve a problem is taken as the fault of the individual researcher. But the loosening of rules that govern the scientific endeavour leads to a consequent 'blurring' of the paradigm and allows the researcher a flexibility unavailable in pre-crisis research (normal science). This new-found flexibility for a researcher will manifest itself in different ways. The most common responses of individuals in the relevant scientific community are: to make *ad hoc* modifications to a theory; to create different versions of the theory in which the unsolved problem is somehow glossed over or removed by definition; to return to and to instigate debate on the fundamentals of the field; or to leave this field of research.

What might these *ad hoc* modifications be? The Latin phrase '*ad hoc*' literally translates as 'for this special object or situation'. In the context of scientific theories, an *ad hoc* modification is one made to cover (or otherwise explain in the barest of ways) a rather obvious and potentially damning point of difficulty. No reason is usually ascribed for inclusion of such a modification other than its fixing the troubled area of theory. Most *ad hoc* measures cannot be experimentally tested. One often quoted example of an *ad hoc* modification to a theory occurred in the Chemical Revolution. The problem in this case was that when metals were heated to form what was called a calx, the calx weighed more than the original metal. According to phlogistic theory, the calx should have weighed less because phlogiston was supposed to have been given off in the process of heating the metal. In order to account for the observed weight gain, some advocates of the phlogiston theory offered the explanation that phlogiston must have negative weight![41] This suggested remedy to the problem of weight gain on calxification was not empirically testable.

Kuhn lists some of the most frequent reactions to crisis as follows:

> Confronted with anomaly or with crisis, scientists take a different attitude toward existing paradigms, and the nature of their research changes accordingly. The proliferation of competing articulations, the willingness to try anything, the expression of explicit discontent, the recourse to philosophy and to debate over fundamentals, all these are symptoms of a transition from normal to extraordinary research.[42]

The proliferation of theory versions and argument over fundamental concepts which occurs during crisis is similar to a state of pre-paradigm science.[43] Crisis is also a time of much psychological discomfort for the scientists involved. They feel a peculiar kind of professional insecurity. The source of their frustration is the failure of the normal scientific methods to provide acceptable solutions.[44] Kuhn (rather dramatically) labels such insecure feelings in scientists as 'the essential tension'.[45] Figure 3 shows that once a crisis state has begun there are only three ways in which it can be ended. First, the problem can be solved within the paradigm. The efforts of researchers under pressure (and with some of their previous restrictions lifted) can yield benefits. The problem then turns out to have been another scientific puzzle after all, though a most stubborn and difficult one. Extraordinary science gives way to normal science, although it may leave some lasting effects. For instance, the period of extraordinary science may have indicated that formerly taboo procedures are really quite legitimate and these are then incorporated into normal scientific practice. In any case, the problem is placed on the pile of solved puzzles and normal science resumes.

The second way in which a crisis state may be dispelled is for the problem to be dropped into a 'too-hard basket', that is, to be forgotten about for the present. The problem is placed on the shelf where it can await more determined (and perhaps more technically advanced) research endeavours. If this is to occur though, the importance of the problem must have (somehow) diminished. If this were not the case and the problem remained of primary importance, then clearly a continuing effort at solution would be made.

What happens if a strong need for a solution continues and if the anomaly remains unsolved for a sufficiently long period of time? This leads to Kuhn's third and final way of ending crisis in science. It may occur to a researcher engaged in trying to reach a solution that the problem cannot be solved within the context of the reigning paradigm. Kuhn claims that no paradigm can ever solve all scientific problems that appear. This, he reasons, is because scientific research is mainly 'a strenuous and devoted attempt to force nature into the conceptual boxes supplied by professional education'.[46] Nature does not always 'feel obliged' to fit comfortably into such artificial boxes and occasionally the packing attempts fail miserably. When an acute failure occurs a new conceptual box—a new paradigm—is required. The proposal of a candidate for paradigm as a method to end a crisis is the most extreme of Kuhn's three possibilities. Recall that Kuhn states that there is only one reigning (monolithic) paradigm at any one time in a scientific discipline. The overthrowing of a reigning paradigm and the installation of another is what Kuhn calls a scientific revolution.

Before there can be any revolution, there must be a candidate for paradigm proposed that accounts for most of the same phenomena as does the reigning paradigm and, most importantly, solves the anomaly (or anomalies). Otherwise there would be no point in the candidate paradigm being given serious consideration. Note that such a candidate is not an alternate version of already accepted theory. It must be radically different in concept to the existing paradigm. If this were not so, the candidate would not be able to provide a suitable solution.

Does an alternative theory always arise in crisis or is it the cause of crisis? In 1772 Antoine Lavoisier wrote in a sealed, secret note to the secretary of the French Academy that he would bring on a revolution in chemistry. Kuhn claims that in the case of the Chemical Revolution the problem of weight gain on calxification constituted an anomaly for the phlogiston theory and brought about a crisis state. However, a check of the respective dates of events does not bear this out. The symptoms of crisis which Kuhn describes (for example, *ad hoc* modifications to theory, essential tension, similarity to a pre-paradigm state) appear after Lavoisier's theory became known in 1772. If the weight gain problem acted as the catalyst of the crisis state, then one would have expected the symptoms of crisis to be evident earlier. On Kuhn's rendering, Lavoisier's theory should have appeared later than, and as a result of, the crisis in chemistry. It would appear more plausible that it was the advent of Lavoisier's theory and its solution to the weight gain problem that brought on crisis and not the other way round.

The very idea of paradigm change is going to conjure up a large degree of professional resistance (both psychological and social) in the relevant field. Kuhn elaborates:

> professionalization leads . . . to an immense restriction on the scientist's vision and to considerable resistance to paradigm change . . . By ensuring that the paradigm will not be too easily surrendered, resistance guarantees that scientists will not be lightly distracted and that anomalies that lead to paradigm change will penetrate existing knowledge to the core.[47]

In Kuhn's view, the individual who postulates a new candidate will be one who is significantly involved in the crisis. Kuhn's mechanism for a change in scientific allegiance is based on Gestalt psychology. The visual 'Gestalt Switch' is a well-known psychological effect. Consider Figure 4, which shows a wire cube (the Necker Cube). Which vertical face of the cube appears closest? Most people would say the face marked number 1, but some would say the face marked 3. As you look at the Figure the closest face seems to switch back and forth from 1 to 3. (This switch can be made by concentrating first on the bottom left corner of the cube and then concentrating on the top right corner.) This is just one example of a visual Gestalt switch.[48] Kuhn has adopted a similar mechanism for the switching of allegiance from one paradigm to another.

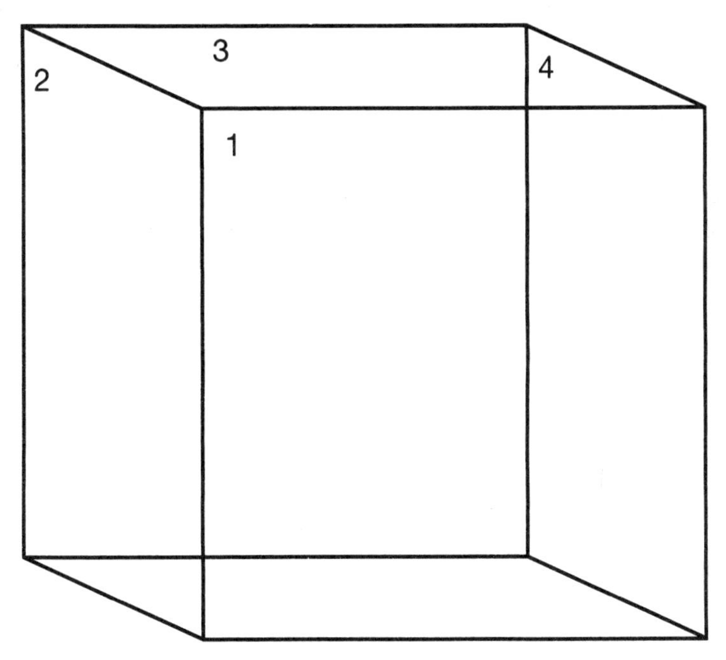

Figure 4 The Necker Cube; which face of the cube is closest, 1 or 3?

When does a scientist resort to the drastic step of proposing a candidate for paradigm? Kuhn's account runs as follows. A scientist in the thick of crisis, having been unable to solve the anomaly within the confines of the reigning paradigm, will suddenly see the problem from a totally new perspective. It is as if the scientist's perception changes from one characterization of the field to another, in much the same manner as perception changes in a visual Gestalt switch. Something will fall into place in the mind of the scientist that will enable the development of a new conceptual framework in which the anomaly finds solution.[49]

What sort of scientist is likely to come up with a candidate for paradigm? Kuhn has a straightforward answer to this question. We saw earlier that scientists tend to continue to work within the paradigm even in times of crisis. We also have noted that there is always much resistance to paradigm change, or even its very suggestion. This is especially the case with those practitioners who have large stakes in the reigning paradigm. Kuhn says that the sort of scientist who proposes a new candidate for paradigm will be one who does not have a large stake in the field and therefore has little to lose but much to gain if successful. Such scientists are usually young or relatively new to the relevant scientific enterprise.[50] One example of this was the young Albert Einstein at age twenty-six. In a modest paper in 1905 titled 'On the Electrodynamics of Moving Bodies', Einstein advanced his 'Special Theory of Relativity'. This paper was to shake the whole of physics to its foundations.[51]

What occurs after a candidate for paradigm is proposed? This question is our next concern.

PARADIGM DEBATE AND REVOLUTION

The proposal of a candidate for paradigm (the theory and methods that solve the anomalies of the established paradigm) brings about what Kuhn terms a 'paradigm debate'. This takes the form of discussion (both verbal and written) about the advantages and disadvantages of the two theoretical alternatives by their respective proponents. At first glance, it may appear that the whole issue over which of the alternative scientific schemes should be the paradigm could be resolved by recourse to empirical evidence. In other words, should it not be the case that experiment is the final arbiter in theoretical disputes? It would be very convenient if this were so; unfortunately the situation is not that simple. According to Kuhn, there are a number of complications involved in the choice between paradigm and candidate with the result that the paradigm debate is less than a fully rational exercise.

The astute reader may have inferred that a majority of scientists involved in a crisis may oppose paradigm change merely because of their personal and professional investments in the established paradigm. Although this is strong stimulus for maintaining a conservative attitude in the face of crisis, it is only one of the contributing factors. Recall for a moment Kuhn's explanation of how a new candidate for paradigm originates. The idea comes to the scientist concerned in an instant, similar to a visual Gestalt switch.[52] This is necessary since the conceptual bases of the established paradigm and of the candidate will be mutually exclusive. The scientist has to experience a Gestalt-like switch in his or her perspective of the discipline. What about the other scientists in the field? The same must be true for them, says Kuhn. They too, once presented with the new theory, can only accept it if they have a Gestalt-like switch in perspective. Kuhn describes such experiences as being like religious conversions:

> The transfer of allegiance from paradigm to paradigm is a conversion experience that cannot be forced. Lifetime resistance, particularly from those whose productive careers have committed them to an older tradition of normal science, is not a violation of scientific standards but an index to the nature of scientific research itself. The source of resistance is the assurance that the older paradigm will ultimately solve all its problems, that nature can be shoved into the [conceptual] box the paradigm provides.[53]

If a Gestalt-like switch has not taken place in a scientist, there are inherent limitations in the perceptions (both intellectual and sensory) of that scientist. For example, after Galileo Galilei announced his telescopic discoveries of lunar craters and the moons of Jupiter in 1610, some adherents to the reigning paradigm of Aristotelian cosmology challenged the validity of his observations. Galileo accepted this challenge and allowed inspection of the astronomical bodies with his telescope. Particular individuals claimed after looking through the telescope that they saw no such bodies![54] Those individuals who looked at astronomical objects through Galileo's telescope but claimed that they did not see them could not see them. Their acceptance of Aristotelian doctrine did not psychologically permit them to perceive phenomena contrary to the doctrines of Aristotelianism. In such cases it is clear that what is 'seen' (the observation) depends on what is believed (the accepted theory). Without the occurrence of a Gestalt-like switch to a new way of thinking about cosmology, these individuals were unable to 'see' that which was so plain to Galileo.

Kuhn claims that paradigm debates cannot be fully rational. The positions of the two sides of the debate are incompatible in the sense that there can be little or no agreement on important issues, such as what constitutes a problem or a solution. The empirical evidence cannot be used to decide the question absolutely since the experimental facts are either defined or interpreted differently by the adherents of the paradigm and candidate. Each side of the debate will be conducted in terms of its own theoretical assumptions and methods. They have no common ground upon which to rationally debate the issues. Paradigm debates are therefore circular and the choice between the reigning paradigm and the candidate cannot be made on the results of normal science. Kuhn writes:

> When paradigms enter, as they must, into a debate about paradigm choice, their role is necessarily circular. Each group uses its own paradigm to argue in that paradigm's defense . . . the status of the circular argument is only that of persuasion. It cannot be made logically compelling for those who refuse to step into the circle.[55]

This incomplete logical contact leads Kuhn to state the paradigm and candidate are not only incompatible, but are also incommensurable. By 'incommensurable' he means that there is no common measure of the fundamental concepts of candidate and paradigm.[56] Let's consider two examples of this incommensurability thesis. In the Chemical Revolution, the phlogiston theory was overthrown by the oxygen theory. The former theory postulated that phlogiston is released into the air during combustion. In the latter theory, it is postulated that oxygen gas is taken out of the air during combustion. There is nothing in the oxygen theory which corresponds directly to phlogiston. In addition, the description of combustion in the phlogiston theory cannot be given in terms of oxygen itself, nor can it be given in terms of other entities that are postulated in the oxygen theory. This has the consequence that there exists no literal means by which explanation in one theory can be transferred to the other.

A second example is drawn from the Einsteinian Revolution. In Einstein's 'special theory of relativity' the mass of a body is dependent on the observer's frame of reference (that is, to his or her state of relative motion) and its mass can also be changed into energy. In the paradigm overthrown by Einstein—the Newtonian paradigm—the mass of a body has a fixed value and is constant for all observers regardless of their reference frame. Although the same name (mass) is used in both paradigms, the respective concepts are incommensurable in the sense that the one does not correspond directly to the other. What is supposedly fixed and unalterable by its very definition cannot be otherwise, simply because an observer adopts another state of motion. An advocate of Newtonian mechanics would have been unable to conceptualize a notion of mass that alters from one frame of reference to another. In the Newtonian paradigm a concept such as 'relativistic mass' would be a contradiction in terms.

The foregoing discussion on paradigm debate and incommensurability is summarized by Kuhn:

> We have already seen several reasons why the proponents of competing paradigms must fail to make complete contact with each other's viewpoints. Collectively these reasons have been described as the incommensurability of the pre- and postrevolutionary normal-scientific traditions, and we need only recapitulate them briefly here. In the first place, the proponents of competing paradigms will often disagree about the list of problems that any candidate for paradigm must resolve. Their standards or their definitions of science are not the same . . . [Secondly] Since new paradigms are born from old ones, they ordinarily incorporate much of the vocabulary and apparatus, both conceptually and manipulative, that the traditional paradigm had previously employed. But they seldom employ these borrowed elements in quite the traditional way. Within the new paradigm, old terms, concepts, and experiments fall into new relationships one with the other. The inevitable result is what we must call, though the term is not quite right, a misunderstanding between the two competing schools . . . Only men who had together undergone or failed to undergo that transformation would be able to discover precisely what they agreed or disagreed about. Communication across the

> revolutionary divide is inevitably partial ... [On to] the third and most fundamental aspect of the incommensurability of competing paradigms. In a sense that I am unable to explicate further, the proponents of competing paradigms practice [sic] their trades in different worlds ... Practicing [sic] in different worlds, the two groups of scientists see different things when they look from the same point in the same direction. Again, this is not to say that they can see anything they please. Both are looking at the world, and what they look at has not changed. But in some areas they see different things, and they see them in different relations one to the other. That is why a law that cannot even be demonstrated to one group of scientists may occasionally seem intuitively obvious to another.[57]

Do scientists really live in different 'worlds'? Is there any communication between scientists who advocate different theories or not? Is paradigm debate circular or is incommensurability only partial? There are cases in the history of science where scientists on opposing sides of a paradigm debate do not seem to have much trouble understanding each other's arguments. The German meteorologist Alfred Lothar Wegener put forward his theory of 'Continental Drift' in 1912. An English language version of Wegener's book *The Origin of Continents and Oceans* appeared in 1915 and found much opposition in Britain and North America where the rival theory of 'Permanentism' was dominant. The responses to Wegener's theory showed quite clearly that his ideas were well understood. Thus total incommensurability (in the sense of scientists 'talking through each other') does not seem supportable as a general thesis. A better case can be made for either partial incommensurability or (in a formal sense) for the 'non-translatability' of theoretical terms from one theory to another. If it were the case for two scientific theories that their terms are not capable of being translated from one theory to the other then the two theories would act as separate scientific languages. This does not, in itself, prevent scientists from speaking both languages.[58]

Paradigm debates are settled in one of two ways. Either the candidate for paradigm wins the debate and a scientific revolution occurs, or the paradigm holds its own and the candidate loses—there is a failed attempt at revolution. What decides which side wins and which loses? Kuhn says that the question is decided by the number of scientists who advocate the reigning paradigm as against the number who advocate the candidate. If, after a period of debate, more researchers support the candidate than the established paradigm in a given scientific field then a revolution will have taken place in this field and the candidate will become the new paradigm.

A scientific revolution, as Kuhn describes it, is not something that can occur quickly since it depends on the conversion of most members of a scientific community. Conversion to a new way of viewing the world not only cannot be forced, it cannot even be logically compelling. Advocates of the candidate in a paradigm debate can only attempt to convert those opposed to change by a process of persuasion which may lead to a change in their perspective of the field. Once the conversion of individuals from one world-view to another is complete, the newly converted scientists can only view their field from its new perspective, according to Kuhn. Thus the convert will be unable to interpret fully the arguments of those who have not switched over to the new world-view. Here the mechanism of conceptual change differs from the visual Gestalt switch. The visual Gestalt switch can be done and then reversed (as with the cube in Figure 4). Unlike the visual Gestalt switch, a change in world-view cannot be undone, because Kuhn claims that the actual mechanism for switching world-views is not itself reversible.[59] If, however, scientists can at least partially understand each other's positions (as in the continental drift case) then such switches, if they occur at all, cannot be irreversible.

Although it may be the most obvious way, conversion is not the only means by which the numbers of supporters for paradigm and candidate can change. New researchers entering the field are more likely to embrace the new world-view (the candidate) and older adherents of the reigning paradigm can (and do) leave the field by retiring from scientific research or simply by dying! If scientists are to remain in research, they cannot cease being advocates of a paradigm without simultaneously changing their support to an alternative. To do otherwise is simply to leave scientific research altogether.[60] The attempt at revolution will be successful when a majority of researchers in the field identify their allegiance with the new world-view. The candidate will then become the new paradigm and those still clinging to the old paradigm will be effectively 'read out' of the field. This occurs when researchers working under the new paradigm take control of the journals in the field and can thereby deny publication to the adherents of the old.[61] There are always some scientists, of course, who never renounce the old paradigm. Joseph Priestley, for example, never really abandoned the phlogiston theory.[62]

Scientists on both sides of a paradigm debate seek the allegiance of the relevant scientific community and it will be this allegiance (in terms of which side commands and maintains a majority of researchers) that will draw the debate to its conclusion.[63] Why do scientists convert to a new world-view if such conversions are not logically compelling? Kuhn has no general answer for this question. He asserts that a number of factors come into play and, in any case, each crisis situation will be different, although there will be factors common to each crisis. The most important of these are: that there is always a very strong urge to resolve a crisis situation (and ease 'the essential tension'); the extent to which the candidate can solve the anomalies; and the aesthetic appeal of the triumphant theory. Kuhn states:

> Individual scientists embrace a new paradigm for all sorts of reasons and usually for several at once . . . Probably the single most prevalent claim advanced by the proponents of a new paradigm is that they can solve the problems that have led the old one into crisis. When it can legitimately be made, this claim is often the most effective one possible.[64]

If, instead of victory, the candidate fails to muster sufficient support in the relevant scientific community, then the outcome of the paradigm debate will be a failed attempt at revolution. Normal science under the reigning paradigm will return in this case. There may be other challenges, other crises in the future of this field, or there may not. Normal science may continue indefinitely, as can be seen in Figure 3 (p. 23). One example of a failed revolution was the attempt by Alfred Wegener to overthrow the reigning paradigm in geology of 'Permanentism', in favour of his version of 'Continental Drift'. The verdict of the geological community at the time was to retain permanentism.[65]

POST-REVOLUTION

After a successful scientific revolution, a new normal science will emerge under the new paradigm. The description of science as a paradigm-guided activity that aims to solve puzzles will again be applicable, although the approach, the methods and the choice of puzzles to solve will mostly differ.[66] There is much mopping up to be done after a scientific revolution. There will be many things overlapping the old and new paradigms, such as uninterpreted raw data, scientific instruments, experimental procedures, and so on. These have to be reinterpreted or explained within the theoretical framework provided by the new paradigm. There will also be a need for textbooks to be written to instil the new paradigm into the next generation of researchers. We noted earlier that textbooks provide exemplars of normal scientific practice. This tradition will be continued in any new paradigm.[67]

Another curious thing happens after a scientific revolution. Once a revolution is successfully concluded, the textbooks are rewritten in a manner such that a new student might be forgiven for not ever knowing that a revolution had taken place. Kuhn calls this phenomenon 'the invisibility of revolutions':

> [the textbooks] have to be rewritten in whole or in part whenever the language, problem-structure, or standards of normal science change. In short, they have to be rewritten in the aftermath of each scientific revolution, and, once rewritten, they inevitably disguise not only the role but the very existence of the revolutions that produced them.[68]

A parallel can be drawn with events depicted in Orwell's classic novel, *Nineteen Eighty-Four*. This story is set in England but it is an England which forms part of the state of Oceania. The government of the day is totally repressive and authoritarian. This government controls every aspect of society by controlling what its citizens think. When it is politically expedient for the government, it has the official history of the country rewritten. For example, if at one time the country is at war with its eastern neighbour and allied with its western neighbour and then the situation is reversed, the history is rewritten to show that they have always been at war with their western neighbour and allied with their eastern. The following extract describes this situation:

> The Party said that Oceania had never been in alliance with Eurasia. He . . . knew that Oceania had been in alliance with Eurasia as short a time as four years ago. But where did that knowledge exist? . . . And if all others accepted the lie which the Party imposed—if all records told the same tale—then the lie passed into history and became truth.[69]

A more relevant example is an average text on classical mechanics. Such a textbook tends to introduce the relativistic correction formulae for changes in mass, length and time at speeds close to the speed of light without necessarily mentioning the Einsteinian Revolution in which these formulae were accepted. Indeed, some textbooks seem to give the impression that these relativistic formulae follow (in some way) from Newtonian mechanics! It is by such methods that scientific revolutions are made invisible.

Normal science is a time when there is much progress in the sense of accumulation of knowledge within a given paradigm. This is one of the two forms of progress identifiable in Kuhn's theory. The other form of progress is progress through revolution. A new paradigm solves the important outstanding problems of the old paradigm (at least as far as the adherents of the new paradigm are concerned).[70] This form of progress is from one conceptual basis to another and as such there will be losses as well as gains. Whatever the losses, the benefit of the gains outweighs them. These losses are sometimes individually or collectively referred to as 'Kuhn Loss'. No paradigm ever solves all its problems.

A good example of Kuhn loss appeared in the Chemical Revolution. Musgrave details the relevant sequence of events:

> In 1772 Lavoisier burned sulphur and phosphorus in air confined over water, and noted the reduction in the volume of the air and the increase in weight of the sulphur and phosphorus. The experiment was not new, but Lavoisier's interpretation of it (contained in three sealed notes deposited with the secretary of the French Academy) was.

> In the first note (dated 10 September 1772) Lavoisier states that when phosphorus burns, air is absorbed. In the second note (dated 20 October 1772) he remembers the phlogiston theory, and says that as phosphorus releases phlogiston it absorbs air. Finally, in the third note (dated 1 November 1772) he takes the bold, though apparently obvious step of dispensing with phlogiston.[71]

This description indicates how Lavoisier did away with the entity phlogiston as an essential part of his explanation of the process of combustion. Once removed, problems associated with phlogiston disappeared but so did some adequate explanations of chemical phenomena. Using the phlogiston theory Joseph Priestley was able to explain the release of 'fixed-air' (carbon dioxide) on heating metals, whereas oxygen theorists could explain the release of 'fixed-air' only if charcoal was present during the heating process.[72] Any problems that are solvable in the old paradigm but not in the new must appear to be of little or no consequence within the new world-view.

The new paradigm will solve puzzles that have gained importance within that paradigm. Decisions about which problems are important and which are not will be dependent on what worth (or value) is attached to finding a solution. Such decisions, clearly, are value judgements made by the researchers concerned. These decisions vary and depend on such factors as socio-economic needs, background knowledge, religious and other beliefs. (Historical case-studies have indicated this particularly well.) Thus scientific values as well as scientific standards change from paradigm to paradigm. Progress through revolution generally results in a net growth over time of the content of the paradigm over its predecessor; there will generally be more puzzles solved in a paradigm's lifetime than its predecessor solved. However, it needs to be stated that progress in Kuhn's scheme of the workings of science is not progress towards any special goal, such as truth about the external world.[73]

Kuhn's emphasis on purely empirical matters leaves much of the scientific endeavour unaccounted for. Non-empirical problems (those of a conceptual nature) are concerned with the theoretical consistency, coherence and clarity of theories. These considerations hold equal importance with empirical problems, especially when a choice is being made between rival paradigms (or theories).

RATIONALITY AND DEMARCATION OF SCIENCE IN KUHN'S THEORY

'Rational' is a much-used word in philosophy, as is the term 'rationality'. What does it mean to say that an action is rational? A basic sense of rational action is as follows. An action is rational if the individual performing the action has reasons for believing that the action will bring about (or assist in bringing to fruition) his or her aim (whatever that aim may be). To hold some reasons yet act against them is irrational—such an action goes against reason.

If, for example, I wish to travel to Bali as directly as possible then I would take a particular route because it is the shortest. The action of taking this route is fully rational since it will most effectively achieve my aim. If I wished to get to Bali but took a route that did not go to Bali and I did this knowingly, then this action would clearly be irrational. Suppose again that I still wish to travel to Bali as directly as possible but do not take the shortest route. Instead I end up taking some slightly longer route which will still get me to Bali. I do so because as I was about to embark on the shortest route I encountered someone travelling the same route who was wearing a red shirt and discovered that I detest red clothes. Such a course of action is not fully rational since it does not conform to the most efficient way to achieve my aim. However, although this action is not fully rational, it is not irrational either since there are reasons for my doing it (although, in this example, highly subjective ones). In general then, an action may be fully rational or may be not fully rational (that is, the action may be a-rational) depending on what sort of reasons are held by the individual who wishes to achieve an aim. (Note the distinction made here between being a-rational and being irrational.)

What about the implications of rationality in Kuhn's theory? Kuhn readily admits that there are no logically compelling reasons for transferring allegiance from one paradigm to another. Thus it is not possible to specify objective criteria, within Kuhn's theory of science, for making a fully rational choice between paradigm and candidate. In other words, there cannot exist a set of rules (or an 'algorithm') that determines the scientist's action. Kuhn's opinion is based on his arguments for the applicability of an incommensurability thesis. If, however, Kuhn is wrong on the matter of total incommensurability then this also places in question his conclusion regarding the lack of rational choice between paradigm and candidate. The choice between competing worldviews is not a fully rational one but neither is the choice necessarily irrational. The choice need not be irrational because there are reasons why scientists do switch their allegiance, as we have already noted.

Many of Kuhn's comments imply (explicitly or implicitly) that extra-scientific factors play important roles in the conduct of scientific research. These factors have a strong bearing on the question of whether there can be any form of rational choice between paradigm and candidate. Examples of these comments by Kuhn can readily be cited: a scientist does not question the paradigm because this behaviour will be frowned upon by other scientists; the inability to solve particular serious problems produces 'tension' and 'crisis'; scientists experience a Gestalt-like conversion similar to a religious experience; the scientist who advances a candidate for paradigm is usually new to the field and does so because he or she has a smaller personal stake at risk than an established researcher; or paradigm change depends on persuasion of scientists rather than rational argument. Despite these and many other references by Kuhn, he provides no real analysis of the social and psychological implications of the issues confronted in scientific research.[74] Kuhn also, aside from the mere mention of the effects of non-scientific beliefs on scientific endeavours (for example, philosophical or religious convictions), fails to draw any firm conclusions about the relationships between scientific and non-scientific beliefs.

We turn now to the question of how the distinction between science and non-science might be specified in Kuhn's theory. The traditional features that have been used to segregate science from other human endeavours—the (presumed) certainty of science; the impartiality of experimental facts; the (assumed) objectivity of scientific method and so on—are either absent or are 'blurred' in Kuhn's account of science. It follows then that the distinction between what is science and what is not must also be blurred. In other words, it appears that Kuhn cannot draw a fixed and definite line of demarcation between science and non-science.

If there is no explicit demarcation line, is it correct to say that there is some form of indeterminacy (or grey area) in distinguishing between science and non-science? On one hand, it seems obvious that disciplines such as physics and chemistry should be designated as 'real' sciences. On the other hand, it also seems plain that such disciplines as literature and law are not sciences. What about those disciplines that occupy the middle ground between what clearly seems to be science and what does not? Many researchers working within these middle disciplines—for example, sociology, cultural anthropology, parapsychology—think of their field as a science. (Whether or not it is correct to call such disciplines a science or not has little to do with whether it is so labelled by its researchers. The general prestige accorded to science leads many of them to embrace the name of 'social science' for their fields.) How are disciplines in the blurred area between science and non-science to be characterized in Kuhnian terms? If such disciplines could be so characterized, by what method could this be achieved? These questions are not easily answered in Kuhn's theory.

It is tempting to give as an initial response the answer that Kuhn's account of science cannot rule out any field as a candidate for paradigm (mature science) status, but that most disciplines are in a pre-paradigm stage. Recall that Kuhn describes pre-paradigm science as that stage in the development of a discipline where there is no common set of guiding principles for the practitioners of the field. The discipline of history, for example, is a field of research that seems to fit this description. Disagreement is rife amongst groups of historians, who constantly debate the significance and interpretation of historical data. We can add one further point about pre-paradigm science. Kuhn would readily acknowledge that science is an empirical enterprise; that is, scientific theories are tested by experiment. If a field has no empirical input from the natural world (as in law) then it can hardly be characterized as a science. (Mathematics is not empirical either, but one finds its natural expression in the form of the equations of physical science and this characteristic is not shared by any other non-empirical field.)

Since many of the traditional features of science are claimed to be absent or sufficiently blurred, can we point to anything in Kuhn's account that remains as a hallmark of science? Kuhn defines (mature) science as a paradigm-guided activity but fields such as economics have researchers who claim to be working within some paradigm or other. According to Kuhn, if a field is of paradigm status, then there must, at least, be a high degree of consensus in that field as this is a prerequisite to formation of a paradigm. In order to explore the claim of paradigm status made by some social scientists, we shall consider an example drawn from the field of economics since this is claimed to be the discipline with the greatest amount of consensus in the social sciences.[75]

The basic theoretical ideas contained in Keynes's well-known treatise, *The General Theory of Employment, Interest and Money*, are straightforward. In brief, the theory claims that the value of everything produced in a capitalist economy during a period is equal to the total of all incomes received in that period. Circular money flow from business to public (rent, wages, interest and so on) and back to business (the purchase of goods and services) does not happen automatically and there are leakages in the flow (savings, taxes, imports). These leakages can be offset by spending injections such as export sales, borrowing and government expenditure. If these injections are just as large as the leakages, then spending equals the value of production; everything that has been produced can be sold and there will be material prosperity. The theory also states that this process cannot continue uninterrupted for long. Investment will eventually fall short of savings and expenditure will fall short of the value of goods and services produced. This results in a reduction of production, a decline in incomes and less employment. This continues until the economy stabilizes at a low level of income equilibrium with high unemployment and large unused productive capacity.

When Keynes first proposed his theory, there was little disagreement between economists on its essential features. Indeed this was, in part, because earlier prominent economic theorists (notably Marx and Hobson) had said much the same.[76] A major claim of the theory was that the relation of savings to income would lead to a depressed but stable economy with accompanying high unemployment. In the period 1936–40, economists heatedly debated Keynes's theory, but did not reach any general consensus. The post-depression years 1942–47 were a test period for Keynesian economics. Looking back on those years leads most economists to the conclusion that Keynes's theory is correct in its major claims and general approach (although they may and do disagree on the finer points).[77] Thus by reference to historical information, economists found common ground by which some level of consensus could be reached.

However, the more specific one tries to be regarding the theoretical bases of social science theories, the smaller the degree of consensus that is achieved. The primary reasons for this can be seen in the above example and are as follows:

- not all of the premises (basic statements) of the field are agreed on (usually only the 'matters-of-fact');
- premises that are agreed on can be subject to differing interpretations, which may and usually do lead to different conclusions;
- the testing of social science hypotheses or conjectures is usually either not possible in practice or only possible in historical perspective—with hindsight.

Do these three inferences offer a method to separate the 'real' sciences from the (so-called) social sciences? At a first glance they do look attractive, provided that researchers in the social sciences do not claim that their fields are actually sciences in a pre-paradigm stage. If we are to use the above three inferences as a separating method, then we need to establish if the 'real' sciences themselves do indeed have a high level of consensus (as is mostly taken for granted). Could a level of consensus significantly less than 100 percent be sufficient in some scientific paradigms? If this could be shown to be the case, it would tend to undermine the use of the above three factors for the purposes of demarcation.

Consider the state of the physics discipline of mechanics in the late seventeenth and early eighteenth centuries. If one accepts

Kuhn's theory, then it follows that there was and could be only one dominant paradigm of mechanics at any one time. In 1686 the German philosopher and mathematician Gottfried Wilhelm Leibniz published an article which was to set off a long and bitter debate now referred to as the 'Vis viva Controversy'.[78] This controversy was about which quantity, mv (mass multiplied by velocity) or mv^2 (called vis viva meaning 'living force'), was the correct measure of the force of a moving body. This episode in science became far more than a mere debate when the British Newtonians joined with those advocating the quantity mv in 1717. Let's now consider now some of the relevant and important aspects of this case.

In 1718 Giovanni Poleni, the professor of physics and astronomy at the University of Padua, Italy, conducted free-fall experiments to test which quantity was the correct measure. William s'Gravesande had argued for the Newtonian view in his *Mathematical Elements of Natural Philosophy* in 1719. He subsequently changed his opinion after performing similar experiments.[79]

The experiments of Poleni and s'Gravesande incited a series of counter-experiments and arguments by British Newtonians over the years 1722–28. This only resulted in an interpretation of the experiments which fitted their opinion.[80]

In 1743 the French mathematician Jean D'Alembert claimed in a published work that the whole controversy was merely 'une dispute de mots' (a dispute over words). D'Alembert argued that the effects of a force could be measured in two different ways: as mv and mv^2. Yet D'Alembert's remarks did not result in consensus between the two groups and therefore did not end the controversy.[81]

The experimentation of Poleni and s'Gravesande is an example of empirical evidence being used to arbitrate between different views. Empirical sciences have this resource for settling certain kinds of disputes—although not all—due to the theory-ladenness of observation and to underdetermination by the available data. At some later time, if the facts were to be agreed upon and more evidence be forthcoming, then the disagreement may well be dissolved.

This brings us to the second point. Both theoretical preconceptions (theory-ladenness) and the state of knowledge about matter itself during the 'vis viva' controversy were major factors in it enduring so long. Leibniz, for example, had been attacked particularly about the apparent non-conservation of 'vis viva' in (inelastic) collisions. He answered that it was not lost to the universe, but that any dissipated 'vis viva' went into the small parts of a body's matter,[82] though Leibniz had no way of substantiating this claim. Henry Pemberton, a friend and colleague of Newton, published what he called a refutation of Poleni's work in which all he had actually done was to show how to get the result using **mv** (momentum) considerations.[83]

This, in turn, brings us to the third point. D'Alembert's paper did not resolve anything. After 1743 the controversy appeared to die out to many observers but only because the Newtonians thought that they had won and, in any case, they were not prepared to listen to alternative views. There was no decisive winner. Neither group could gather sufficient evidence or arguments to trump the other. Consensus in this case was unattainable until further theoretical advances were achieved and more empirical findings came to light.

The case of the 'vis viva' controversy shows that if there is only one paradigm at any given time then paradigm-guided research cannot require total consensus on all relevant issues. What then, it may legitimately be asked, is an acceptable level of consensus for a field to make the transition from pre-paradigm to paradigm status? Economics, for one, is a discipline with a relatively high degree of consensus but which is not considered a 'real' science. It would also appear that despite the immediate experimental resources available to the 'real' sciences by which some disagreements may be resolved, this method is not always available. When they are available, experimental methods may simply fail to be a successful arbitrator of differences and consequently fail to bring about consensus. Therefore it would appear that within Kuhn's theory, science cannot be consistently demarcated from other human endeavours by appealing to its level of consensus or to the means by which consensus might possibly be reached. It can also be inferred from the above considerations, together with the finding that scientific values change from paradigm to paradigm, that scientific knowledge in Kuhn's account is not much different from other, non-scientific forms of knowledge.

Kuhn claims that the aim of science is problem solving. We have now concluded that it does not seem possible to draw consistently a fixed line of demarcation between science and non-science in Kuhn's theory. Kuhn's specification of problem solving for *the* aim of science only exacerbates this situation, for this aim is so wide that it could (in the opinion of some commentators) include such diverse fields as Oxford philosophy or even organized crime![84]

3
Lakatos's Methodology of Scientific Research Programmes

Imre Lakatos (1922–74) was an outstanding contributor to both the philosophy of mathematics and the philosophy of science. He spent most of his academic career at the London School of Economics (LSE), having fled from his native Hungary after the 1956 uprising in that country. Lakatos was in good philosophical company at the LSE with Sir Karl Popper on staff. Lakatos criticized the theories of both Kuhn and Popper, but it was Popper's philosophy of science that he extended.

Lakatos published his ideas on scientific methodology in 1968. In the interval 1968–70 he revised and expanded his initial ideas and republished.[1] What was Lakatos's motivation to postulate yet another philosophical account of science? After all, Popper's views on philosophy of science had been well received (especially at the LSE) and older schools of philosophy of science still retained minority followings. Influences from the other side of the Atlantic were making an increasing impact too, especially Thomas Kuhn's *The Structure of Scientific Revolutions*. In the wake of Kuhn's theory, sociologists realized that the processes governing the generation of scientific knowledge were an area which they could legitimately explore.

Lakatos, like Popper and Kuhn, saw no future in the efforts of those philosophers whom he collectively labelled as 'justificationists'. These included the 'Logical Empiricists' and 'Classical Rationalists' (such as the German philosopher Immanuel Kant).[2] The logical empiricists needed an empirical basis which was certain as a foundation on which to build up theories by use of the process of induction. Classical rationalists, on the other hand, needed a collection of *a priori* principles (principles knowable to the intellect independent of experiment) from which they could arrive at conclusions deductively. In Lakatos's view neither approach could be successful:

> both [types of justificationists] were defeated: Kantians by non-Euclidean geometry and by non-Newtonian physics, and empiricists by the logical impossibility of establishing an empirical basis . . . and of establishing an inductive logic (no logic can infallibly increase content).[3]

Despite his rejection of justificationist positions, Lakatos was very concerned to characterize science as a rational enterprise and to demarcate science from other human endeavours. He saw theories such as Kuhn's as not only erroneous but also somewhat dishonest renderings of scientific activity in terms of social psychology.[4]

Lakatos developed his own theory of science as a partial response to descriptions of science in primarily psychological and/or social terms. Lakatos depicted science as an eminently rational and, in his terms, an honest enterprise. In addition to his conviction that science should not be thought of as an irrational human endeavour, Lakatos was also very much concerned to show that a clear distinction could (and should) be made between science and non-science (or 'pseudo-science' as he preferred to call it). He thought it was of great importance to both the freedom of the individual and to society in general that this distinction be made explicit and unambiguous:

> The problem of demarcation between science and pseudoscience has grave implications also for the institutionalization of criticism. Copernicus's theory was banned by the Catholic Church in 1616 because it was said to be pseudoscience . . . The Central Committee of the Soviet Communist Party in 1949 declared Mendelian genetics pseudoscientific and had its advocates . . . killed in concentration camps . . . the West also exercises the right to deny freedom of speech to what it regards as pseudoscience . . . All these judgements were inevitably based on some sort of demarcation criterion. This is why the problem of demarcation between science and pseudoscience is not a pseudo-problem of armchair philosophers: it has grave ethical and political implications.[5]

Lakatos found a starting point for his own theory in the writings of Popper. Much of Popper's 'Method of Conjecture and Refutation' (also referred to as Popper's version of 'Sophisticated Falsificationism') appeared to be on the right track but Lakatos obviously thought that Popper had not taken his programme to its logical conclusion. Lakatos began with Popper's sophisticated falsificationism and developed it into the 'Methodology of Scientific Research Programmes'. Note that both Popper's and Lakatos's accounts of science are prescriptive (or normative)—ones that detail what scientists should do—as against Kuhn's theory, for example, which is primarily a descriptive account—one which describes what scientists did.

CONVENTIONALITY AND THE EMPIRICAL BASIS

Lakatos accepted that all observations are theory-laden; that nature does not provide a strict dividing line between 'fact' and 'theory'.[6] Theoretical aspects are always a component of observation. This is why there cannot be a firm, undisputable empirical basis for science. If there is no clear separation between fact and theory then theories cannot be (logically) proved. Perhaps more surprisingly, theories cannot be (logically) disproved either! How can this be the case? Consider what one does in an attempt to refute a theory—one uses some experimental result (or a fact if you like). However, all so-called 'facts' are themselves theory-laden and therefore questionable; pure unadulterated 'facts' do not exist. Since the truth-value of a factual statement cannot be established by the use of logic alone, any clash between a factual statement and theoretical statements only indicates that some aspect of one (or more) of the theoretical statements is not compatible with the factual statement. Lakatos summarizes this position:

> *no factual proposition can ever be proved from an experiment.* Propositions can only be derived from other propositions, they cannot be derived from facts: one cannot prove statements from experiences . . . If factual propositions are unprovable then they are fallible. If they are fallible then clashes between theories and factual propositions are not 'falsifications' but merely inconsistencies. Our imagination may play a greater role in the formulation of 'factual propositions', but they are

both fallible. Thus *we cannot prove theories and we cannot disprove them either.*[7]

Lakatos therefore rejected naive versions of falsificationism, wherein a theory is refuted by an observation that is contrary to the prediction of the theory. In Popper's sophisticated falsificationism, verification of theory predictions is taken to be at least as important as falsification. In this version of falsificationism the appearance of apparent counter-instances does not immediately refute the theory under scrutiny. The impossibility of establishing a firm empirical basis meant that sophisticated falsificationism had to allow for the inclusion of particular conventional elements in scientific theories—elements which are present due to an agreement or decree (*by convention*!). The convention here is that aspects of theory or particular observations are taken to be not refutable, but only because it is so agreed by the researchers concerned. Indeed, some form of methodological decision in these circumstances is necessitated, otherwise there will just be stagnation. Since scientific method is essentially a recipe for the conduct of scientific research, a method that only yields the result that theories are both unprovable and unrefutable is not a scientific method at all! In sophisticated falsificationism, the methodological decision is to include some agreed elements into a scientific theory.

The development of a scientific theory must start somewhere. One way to proceed is to work from some unquestioned (as distinct from undisputable) empirical foundation. Popper expresses this idea through the following analogy:

> Science does not rest upon solid bedrock. The bold structure of its theories rises, as it were, above a swamp. It is like a building erected on piles. The piles are driven down from above into the swamp, but not down to any natural or 'given' base; and if we stop driving the piles deeper, it is not because we have reached firm ground. We simply stop when we are satisfied that the piles are firm enough to carry the structure, at least for the time being.[8]

An empirical basis (of sorts) can be established by making some conventional decisions in regard to what will be accepted as non-refutable. Such a basis is composed of a set of what Lakatos and Popper call 'basic' statements. (The term 'observational' statements—in quotation marks—is also used synonymously.) These statements are unfalsifiable by convention. Their truth-value cannot of course, be established, so their inclusion in the empirical basis occurs through the deliberate choice of the researchers. The statements chosen to be included in the empirical basis must be quite particular, according to Popper. They must be of a singular form rather than a universal one.[9] Singular propositions state only particulars: 'this bee has a yellow sting'. Universal statements cannot be expressed in terms of particulars: 'all bees have stings'. An empirical basis established in this conventional manner must only be viewed as a temporary foundation on which to build scientific theory. Experiment may later show a need to revise the decisions taken when establishing the basis. Theories built upon such a basis are then put to experimental test.

How are the required choices, about which statements should form the basis, made? Lakatos says that scientists decide which statements are suitable by use of some current experimental method.[10] For example, if we wished to test a theory which predicted that hydrogen gas would be released in abundant quantities from leaking drain pipes, an accepted experimental method might be to place a lighted match next to the leaking pipe and observe if an explosion follows. The 'observational' statement derived from this experimental method would read something like:

> An explosion occurred when an ignited, phosphorous-tipped match was placed within one centimetre of a six-centimetre crack in a five-centimetre deep by ten-centimetre wide aluminium drain-pipe leaking water, located in our laboratory fume-cupboard, on Tuesday 1 April, at 0900 hours (local time).

Note that experimental methods, although accepted by researchers when initially shown to be useful, change from time to time. When experimental methods do change, the content of theories and perhaps the empirical basis itself may have to be altered, as implied in Popper's analogy of the foundation piles.

The sophisticated falsificationist also uses successful theories (those whose predictions are well confirmed by observation) in the testing of other less established theories. In this way, a well-confirmed theory may (for the purposes of testing) be temporarily viewed as 'unproblematic background knowledge'.[11] By so viewing them, well-confirmed theories act as extensions of our senses. The field of nuclear physics offers a suitable example. The nucleus of an atom is of extremely small dimensions, making investigations of its constituents and behaviour a highly complex and difficult task. The structure and energy levels of a nucleus can be probed using beams of high-energy particles. Much of what emerges from a nucleus after it collides with these beam particles is recorded as streaks on photographic plates. An untrained observer looking at these plates would literally only see marks on transparent plates—the plates would hold no significance (see p. 19).

In order to draw conclusions from an analysis of such plates, a researcher must first assume that the streaks are caused by chemical reactions which occur when light (or other radiation) strikes the plate. This radiation, in turn, is emitted from a detection device that is assumed to be activated by the passage of certain types of subatomic particles. These assumptions lead a researcher to believe that the streaks on the plates are an accurate record of the paths and interactions of emitted particles. Measurements of distances along tracks and angles between different tracks on the photographic plates then provides data by which the mass, speed, charge, energy and lifetime of most particles produced in an interaction may be calculated. Therefore, in order to analyse these plates, a researcher must take for granted a background of photochemical reaction theory, the theory behind the workings of the particle detector (including its electronic circuitry), atomic collision theory, elementary particle theory, the law of conservation of momentum, the special theory of relativity and so on.

THEORIES, PROBLEMSHIFTS AND FALSIFICATIONS

Lakatos states his principle of scientific honesty (or, more generally, of intellectual honesty) as follows:

> Intellectual honesty does not consist in trying to entrench, or establish one's position by proving (or 'probabilifying') it—intellectual honesty

> consists rather in specifying precisely the conditions under which one is willing to give up one's position.[12]

In keeping with this notion of scientific honesty and his conviction regarding the objectivity of science, Lakatos thought that criteria for distinguishing scientific and non-scientific modification of theory, for acceptance and rejection of theories, as well as for the demarcation of science from pseudo-science must be explicitly spelt out. Further, these criteria must be able to withstand rational reconstructions of the history of science. (This last point will be returned to later in the chapter.) Thus, for Lakatos, the main problems encountered in trying to establish a coherent methodology of science consistent with the criterion of intellectual honesty is how to distinguish between scientific and non-scientific alteration of theory and between rational and non-rational theory change—the change from accepting one theory to accepting another.[13] Consider firstly what Lakatos claims a scientist does when confronted with an anomaly. (An anomaly in the Lakatosian sense is generally taken to be some inconsistency or problem in matching theory to observation, but not necessarily a counterinstance to a theory. Note the difference from Kuhn's definition.) Lakatos claims that anomalies are ignored as a standard operating procedure whilst a theory is successful. He cites Prout's chemical theory of 1815 as an example of scientific research that continued successfully despite being immersed in 'an ocean of anomalies'.[14] Prout had claimed that the atomic weights of all pure chemical elements were whole numbers. There were numerous anomalies for this theory in the form of measured atomic weights that were not whole numbers but instead were mixed fractions. Yet Prout and his followers continued to make progress by ignoring these empirical anomalies. Prout claimed that these measured fractional weights were due to impurities in the analysed compound and consequently the experimental method used to isolate them must be unreliable.[15] Prout was later to show that his criticisms were correct, but this took many years.

Anomalies are a nuisance and sometimes an embarrassment for scientists. Although a researcher may not actually address the anomalies directly (in many cases this would not be practical if there are numerous anomalies) they can be got rid of. The easiest way to remove an anomaly is to reinterpret the theory so that the anomaly is rendered *lawlike*. In other words, the theory is interpreted in a manner such that the anomaly becomes a consequence of the theory itself. This option is not always available and, when exercised, it does not result in any increase in the content of the theory. Another option is to add some auxiliary hypotheses to a theory. This can perform the same function of removing anomalies but with the bonus of increasing content. We might well say that the sophisticated falsificationist 'hangs on to the baby whilst gaining *new* bathwater'!

If, however, we have increased the content of a theory by adding extra hypotheses are we really dealing with the same original theory or with a new one? Considerations of this sort have led to more sophisticated criteria for the falsification of a theory. Lakatos details these criteria:

> For the sophisticated falsificationist a scientific theory *T* is *falsified* if and only if another theory *T′* has been proposed with the following characteristics: (1) *T′* has excess empirical content over *T*: that is, it predicts *novel* facts, that is, facts improbable in the light of, or even forbidden, by *T*; (2) *T′* explains the previous success of *T*, that is, all the unrefuted content of *T* is included (within the limits of observational error) in the content of *T′*; and (3) some of the excess content of *T′* is corroborated.[16]

These criteria will now be examined more closely. Since the methodology of naive falsificationism has been shown to be unsuitable, new standards for elimination (falsification) are needed in which rejection of a theory is done for plainly rational (and honest) reasons. First and foremost, no theory is 'falsified' unless there is another theory to replace it. (We use the word 'falsified' with quotation marks as synonymous with eliminated or rejected since, in a strict sense, no theory can be falsified.) The replacement theory must satisfy the conditions 1–3 in the above quotation. Condition 1 states that the empirical content of the new theory *T′* must be greater than theory *T*. By 'empirical content' Lakatos means information about the world that has the potential to be verified by experiment. Thus from *T* to *T′*, there will be a net growth of empirical content. Furthermore, the new empirical content in theory *T′* must include details of events or circumstances that theory *T* either would have been unlikely to predict or would not have predicted at all. (Hereafter we shall simply refer to the details of the new and unexpected empirical content using Lakatos's term—'novel facts'.)

Condition 2 says that the empirical confirmations of the old theory *T* are explained by the new theory *T′*. Let's consider a familiar example. In the Copernican astronomical system the Sun occupied a central place (that is, the system is heliocentric) and the planets revolved in circular paths about the Sun. In order to obtain accurate matches between the observed planetary positions and ones predicted by his theory, Copernicus found it necessary to place some of the planetary centres of motion on slight eccentrics and some planets on minor epicycles. (An eccentric is where the centre of motion of the planet is not located at the centre of the Sun but is placed a small distance away from it. An epicycle is a small circle centred on the planet's orbital path. If a planet is on an epicycle then the planet revolves around the centre of the epicycle but it is the centre of the epicycle that follows a circular orbit about the Sun. A more complex planetary path is thereby produced by these combined motions.) With the use of planetary eccentrics, by suitable variation of the period in which the planet moved around its epicycle and the period in which the centre of the epicycle moved around the Sun, Copernicus managed to match the observed positions with those predicted with a fair degree of precision.

Johann Kepler made several modifications to Copernicus's system. Kepler did away with planetary epicycles and instead postulated that the paths of the planets were ellipses, not circles. Kepler's resulting astronomical system was heliostatic (with the Sun stationary) but not heliocentric and did away with the troublesome epicycles. Kepler's system accounted for all phenomena predicted by the Copernican system—such as stellar positions, solar and lunar eclipses and planetary retrogression—and did so with greater accuracy. Additionally, in Kepler's system, it is possible to plot the orbital paths of comets (and thereby predict their future positions) based on a finite number of observations. This cannot be done within the Copernican system and therefore constitutes a novel prediction in the Lakatosian sense. Kepler's theory not only contained all the unrefuted empirical content of the Copernican theory, it also predicted novel facts. This shows that the conditions 1 and 2 are satisfied by Kepler's theory. Condition 3 was also satisfied when the predictions of cometary positions were experimentally corroborated.[17] It can legitimately be claimed by the sophisticated falsificationist that Kepler's theory 'falsifies' that of Copernicus, since all three specified conditions are satisfied.

Note here that the term 'prediction' is used by Lakatos to include 'postdiction'—a situation where discovery of experimental facts precedes the postulation of theory from which these facts would follow as a consequence.[18] An empirical fact may have little or no significance when discovered if there is no theory to explain it; the significance may only appear after the advent of suitable theory. Postdiction is as important as prediction in Lakatos's theory although he was not the first to emphasize its importance.

Lakatos points out that not all additions or modifications to a theory can count as being scientific. If we allow changes of any and every variety then science would be little different from, say fairy-tales. If this attitude to theory amendment reigned, where would the logic be in scientific methodology? There are standards by which science is conducted and such standards are, and have been, part of the hallmark of science. Lakatos writes:

> Why aim at falsification at any price? Why not rather impose certain standards on the theoretical adjustments by which one is allowed to save a theory? Indeed, some such standards have been well-known for centuries, and we find them expressed in age-old wisecracks against *ad hoc* explanations, empty prevarications, face-saving, linguistic tricks.[19]

Earlier, we defined *ad hoc* modifications as those made to cover some specific explanatory gap in a theory which had been exposed by experiment and that such modifications cannot usually be made the subject of experimental investigation.

Modifications to theory are unscientific, according to Lakatos, if they are merely *ad hoc*, are evasive or are misleading in what they assert. Examples of adhockery are commonplace in the history of science, such as the phlogiston case. Another example is the two synchronizing conditions for planets in the Ptolemaic astronomical system. These conditions were required to achieve a reasonable agreement between prediction and observation. Yet there is no basis for these conditions other than without them the system will fail to synchronize planetary motions in accord with observed positions. The two conditions were imposed *ad hoc* in order to make the system perform accurately.[20]

We have seen that 'falsification' of a theory does not precede the proposal of a better theory. ('Better' here is defined by fulfilling the conditions set out for 'falsification'.) The evaluation of whether a theory is better than another obviously cannot be conducted by looking at it in isolation. The criteria for 'falsification' and theory acceptance requires that a new theory be weighed up against previously held theories. Any auxiliary assumptions in a new theory need to conform to a scientific standard. Lakatos writes:

> any scientific theory has to be appraised together with its auxiliary hypotheses, initial conditions, etc., and, especially, together with its predecessors so that we may see by what sort of *change* it was brought out. Then, of course, what we appraise is a *series of theories* rather than isolated *theories*.[21]

Let's return to the earlier astronomical example. We saw that the heliocentric system detailed by Copernicus in the seventeenth century was 'falsified' by Kepler's heliostatic system. Kepler's system did so because it accounted for all phenomena predicted by Copernicus but with greater accuracy and it made novel predictions, which were to be later verified. Kepler's system was, in turn, followed by that of Sir Isaac Newton. Newton expounded this system in his famous treatise, *Principia*, in 1687. Newton's system of the world embodied his (mathematical) laws of motion and gravitation and these yielded far more precise predictions than had previously been possible. Some of these predictions were experimentally verified in Newton's lifetime. Many other predictions from Newtonian theory only became possible after advances in telescope and optical technology were made. For example, in the nineteenth century astronomers predicted that a planet existed beyond the orbit of Uranus (that is, the planet Neptune). The assumed gravitational attraction of the postulated planet was held responsible for the observed perturbations from the predicted orbit of Uranus.[22] These orbital perturbations are so small that they cannot be detected except with the use of a powerful telescope. Such a telescope was not available in Newton's time.

It is therefore correct to say that given the stated conditions 1–3 for 'falsification', Newton's planetary theory is better than Kepler's and Kepler's theory is better than Copernicus's. This example demonstrates how a scientific appraisal for the sophisticated falsificationist can be done only with a series of theories, not a single isolated theory. In the course of expounding on the idea of appraising a number of theories, Lakatos provides some further definitions. These definitions make explicit relevant aspects of the process of appraisal. A series of theories is held to be *theoretically progressive* if each theory in the series has greater empirical content than its predecessor. Recall that new empirical content means the prediction of novel facts regardless of whether or not such facts are experimentally verified. (Note also that the mere reinterpretation of theoretical terms adds no new content to a theory.) A theoretically progressive series of theories may, in addition, be *empirically progressive* if some of the novel facts predicted (postdicted) are experimentally verified. (A series of theories may be theoretically progressive without being empirically progressive, but not vice versa.) Lakatos's brand of falsificationism places emphasis on the confirmation of predictions (especially novel facts) rather than 'falsifying' instances. Lakatos holds that the comparison of the empirical contents of successive theories is necessary in order to establish progressiveness. Yet this kind of comparison, if indeed possible at all, would be highly problematic. All attempts to quantify measures of the content of theories have failed.

Lakatos uses the term 'problemshift' in regard to whether a series of theories is progressive in one or both senses defined above. A series of theories being theoretically (or empirically) progressive is taken as synonymous with there being theoretical (or empirical) problemshifts. Lakatos's summary of these points is as follows:

> Let us take a series of theories, T_1, T_2, T_3, . . . where each subsequent theory results from adding auxiliary clauses to (or from semantical reinterpretations of) the previous theory in order to accommodate some anomaly, each theory having at least as much content as the unrefuted content of its predecessor. Let us say that such a series of theories is *theoretically progressive (or 'constitutes a theoretically progressive problemshift')* if each new theory has some excess empirical content over its predecessor, that is, if it predicts some novel, hitherto unexpected fact. Let us say that a theoretically progressive series of theories is also *empirically progressive (or 'constitutes an empirically progressive problemshift')* if some of this excess empirical content is also corroborated, that is, if each new theory leads us to the actual discovery of some *new fact*.[23]

A problemshift constitutes changes in the approach of researchers; for example, in the auxiliary hypotheses they postulate (provided such hypotheses are of a scientific standard) together with any modifications made from theory to theory.

Generally it can be said that a problemshift is progressive if it is empirically progressive—if it is theoretically progressive and some of its predictions are experimentally confirmed. A problemshift that does not conform to this is called *degenerating*. Further, a problemshift is considered scientific if it is at least theoretically progressive, otherwise it is pseudo-scientific.[24] All accounts of science place importance on some notion of scientific progress. It seems to be a fundamental norm of scientific research that it be (in some sense) progressive, at least in the long term. Scientific research that continues without any form of perceived progress is destined to fold. In Lakatos's account too, we find a definition of progress, or more precisely the rate of scientific progress. The extent to which a series of theories leads to the experimental verification of theoretical predictions is Lakatos's measure of the rate of scientific progress.[25] In other words, the rate of scientific progress is given by the number of verifications of novel facts in a given time interval.

RECASTING SOPHISTICATED FALSIFICATIONISM AS THE METHODOLOGY OF SCIENTIFIC RESEARCH PROGRAMMES

We have reached a stage in our discussion where Lakatos's structured account of science may be presented. This account is one where science is not just composed of individual theories but is rather a more complex structure in which theories are components. There are similarities between Lakatos's structured account of science and Kuhn's theory. In particular, the Lakatosian 'Research Programme' bears a strong resemblance to a Kuhnian paradigm. This is really not surprising since both Kuhn and Lakatos give holistic explanations of science and scientific change. However, many of the details of their respective theories of science are quite different. In particular, we may initially point to Lakatos's move in distinguishing between the overall holistic structure (the 'Research Programme') and its component theories. (This distinction was not made by Popper[26] and was not explicit in the 1962 version of Kuhn's theory.)

A 'Research Programme' consists of two major structural parts called the 'Hard Core' and the 'Protective Belt'. It is these parts that provide the continuity for any series of scientific theories that emerge from a research programme. There are also two main methodological rules which operate in any given research programme called the 'Positive Heuristic' and the 'Negative Heuristic'. (Note that 'heuristic' literally means serving to stimulate investigation or the process of finding out.) Figure 5 is a flow chart depicting the operation of these two heuristic principles. The hard core of a research programme contains the fundamental tenets of the programme and these (by convention) cannot be questioned. In other words, the adherents of a particular research programme agree (perhaps only tacitly) that the core assumptions of the programme are taken as 'irrefutable'.[27] If a researcher cannot accept a research programme's hard core *in toto*, then the researcher is choosing to opt out of that research programme.

One example of a hard-core tenet of a research programme is the theoretical assumption of drifting continents. Alfred Wegener is generally acknowledged as the first to publish a comprehensive theory of continental drift in 1912, as distinct from earlier vague hypotheses on the matter. Over the next fifty-five years,

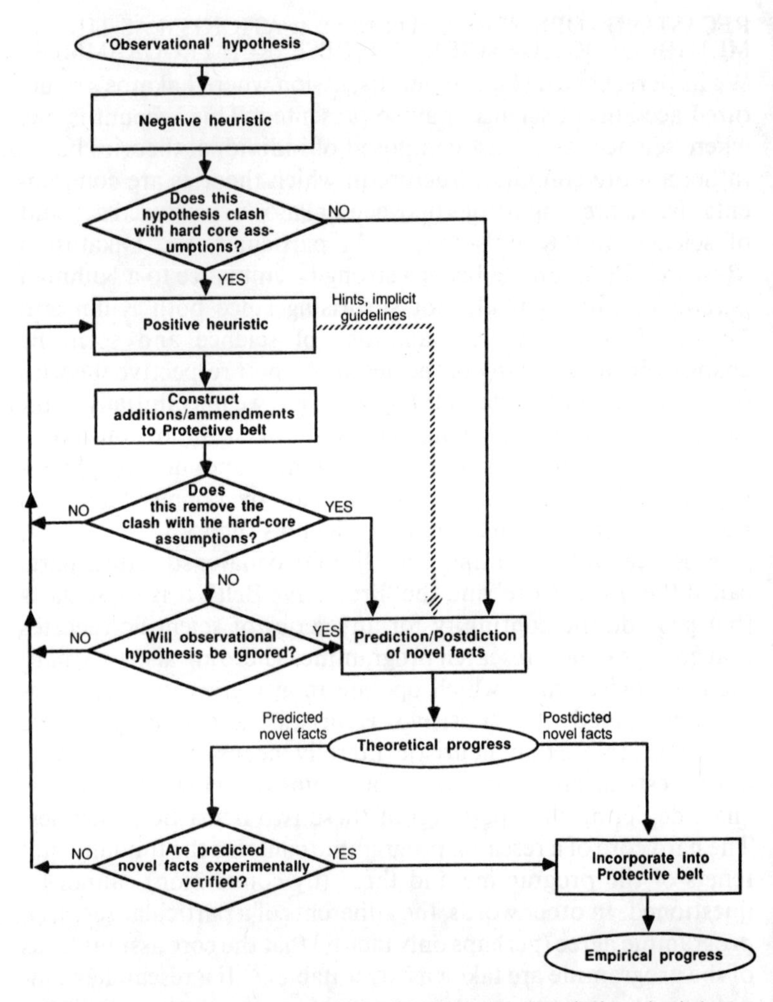

Figure 5 The basic operation of Lakatosian heuristic principles

several different versions of drift theory were proposed by others: A. Holmes, H. Hess, J. T. Wilson, F. Vine and D. Matthews.[28] Each version (or individual theory) can be characterized as forming part of a continental drift research programme. The hard core of this programme would have to include the central assumption that the continents moved laterally over the Earth during geological-time periods. This assumption would be common to all versions of drift theory and could not be doubted by adherents to this research programme.

The origin of a research programme's hard core is not of concern to Lakatos. He argues that once a methodological decision is made by scientists in relation to what constitutes a particular research programme's hard core, then that hard core is thereafter fixed and unalterable. Lakatos only hints at how researchers go about choosing the hard-core assumptions of their research programme. Since these assumptions are fundamental to the establishment of any research programme, Lakatos should explicitly describe the methods by which researchers choose their hard-core assumptions.

The second major structural feature of the Lakatosian 'Research Programme' is its 'Protective Belt'. This consists of the content of the programme other than that making up the hard-core and the two heuristic principles. Lakatos labels the information making up the protective belt 'auxiliary hypotheses', which, unlike the hard-core assumptions, are subject to change. However the protective belt also includes information such as initial conditions—conditions that specify some set of starting parameters for a theory. In Newtonian mechanics, for example, the time taken for an object to fall a given distance can be precisely predicted but this will be possible only if the initial conditions relevant to the state of the object are specified. These initial conditions are the body's starting velocity and the acceleration due to gravity in the vicinity of the body. Lakatos calls the belt 'protective' because it shields the hard core from potentially damaging 'observational' hypotheses. (The term 'observational' hypothesis is used to mean what might otherwise be called an experimental result or just an observation. The explicit use of the word 'hypothesis' indicates its theory-ladenness.)

Let us now consider the first of a research programme's methodological principles—the negative heuristic. The negative heuristic is a rule that tells the researcher what *not* to do, namely, not to modify the hard core of a research programme. Any changes made to a research programme are made by modifying or adding to the auxiliary hypotheses in the protective belt. When an observational hypothesis is found to be in conflict with predictions made by the research programme (that is, there is the appearance of an anomaly) this conflict indicates that there is inconsistency between the observational hypothesis and one or more of the hypotheses contained in the research programme.[29]

How can such inconsistencies be dealt with when they appear? In the propositional calculus of formal (deductive) logic, there is a rule of negation by which an assumption can be refuted by another proposition. This rule is called the '*Modus Tollens*'.[30] The application of this rule to scientific situations can be seen by way of an example.

Assumption: If the orbital paths of all planets are circles then the planet Mars moves along a circular path.
Observational proposition: Mars moves along an elliptical path.
Conclusion: Not all planets move along circular paths.

The antecedent statement of the assumption in this example has been negated. It was made the 'target' of the *modus tollens*. Suppose that there are a number of assumptions from which we draw a conclusion and this conclusion is found to be inconsistent with a proposition which is held to be the case, for example, an observational hypothesis. Then the operation of *modus tollens* does not single out any single assumption (or conjunction of assumptions) to negate. A choice must be made on a basis other than formal logic in regard to which assumption(s) is/are to be negated.

The case of a research programme being tested is much the same. If an observational hypothesis is in conflict with the predictions of a research programme, this implies that an operation of the *modus tollens* should be applied, but at what in the programme should the *modus tollens* be aimed? It is the role of the two heuristic rules to indicate to the researcher where the *modus tollens* should be directed. The negative heuristic gives the straightforward instruction that the hard core is not the 'target' of *modus tollens*.[31] This leaves only the positive heuristic to 'instruct' the researcher where to direct the *modus tollens*. What does the positive heuristic assert? A highly specific answer cannot be given to this question; only a more general response is possible. The positive heuristic of a research programme offers the researcher in that programme a guide to conducting research and constructing the protective belt. The *modus tollens* is directed to the auxiliary hypotheses in the protective belt, as these hypotheses may legitimately be altered. The auxiliary hypotheses are adjusted or replaced by researchers in order to protect the hard core.

This guide to conducting research (the positive heuristic) will usually vary from research programme to research programme. Since the hard core of a programme cannot be changed, the positive heuristic can only suggest ways in which the auxiliary assumptions may be modified or supplemented and thereby develop the research programme. The positive heuristic also saves the researcher from being troubled by the presence of anomalies as long as the programme continues to have some of its novel facts verified. The researcher, guided by the positive heuristic, concentrates on developing the research programme and not on solving its anomalies.[32]

How does this development of the research programme occur? The positive heuristic shows 'avenues' by which a researcher may construct 'models' that describe the particular phenomenon under study. Lakatos defines a model and its relation to the positive heuristic as follows:

> A '*model*' is a set of initial conditions (possibly together with some of the observational theories) which one knows is *bound* to be replaced during the further development of the programme . . . the positive heuristic is there as a strategy both for predicting (producing) and digesting them.[33]

A researcher attempts to improve each subsequent model by making modifications and/or additions, thereby generating a succession of more and more complicated models. A research programme develops by this process and each member of the series of theories within the programme will embody a particular model. Lakatos illustrates this aspect of a research programme using the development of Newtonian celestial mechanics as an example. Newton's first model of our planetary system had a fixed point-like Sun and one point-like planet. This model was amended by including more planets, all revolving around a common centre of gravity. The next modifications were to make the Sun and planets spatially extended bodies. The planets and the Sun are also spinning bodies, so this too was incorporated into another model and so on. Each of these modifications solved a puzzle (of sorts) and each was to be expected:

> Most, if not all, Newtonian 'puzzles', leading to a series of new variants superseding each other, were forseeable [sic] at the time of Newton's first naive model and no doubt Newton and his colleagues *did* forsee [sic] them: Newton must have been fully aware of the blatant falsity of his first variants. Nothing shows the existence of a positive heuristic of a research programme clearer than this fact: this is why one speaks of 'models' in research programmes.[34]

We can now point to a number of similarities (and some dissimilarities) between the theories of Lakatos and Kuhn: both are holistic accounts of science; both postulate steady stages of research where the guidelines for that research are either implicit or partially articulated and both theories require researchers to continue working within their accepted scientific norms when troublesome problems appear. However, the ways in which the theories of Kuhn and Lakatos deal with anomalies show one of the principal differences between them. In Kuhn's theory, a persistent problem may take on the status of a counter-instance (that is, a Kuhnian anomaly) if other crucial factors are brought to bear. An anomaly, for Lakatos, can only be considered as a counter-instance with the value of hindsight and after the advent of a new, successive theory.[35] Lakatos would also say that the history of science is best characterized as a history of competing research programmes. He points out (contra Kuhn) that usually there is not a single paradigm dominating a field at any one time but a plurality of research programmes all in competition.[36]

The hard-core assumptions of a research programme are supposed to be unfalsifiable by convention. Yet if we look at the history of science we find that there have been research programmes in which core assumptions have been changed and/or falsified. One example of a changing core is found in the continental drift research programme. The core of Wegener's theory originally included the assumption that continents had ploughed through the more dense ocean floors. This assumption is absent in later versions of drift theory, replaced by alternative assumptions.[37]

Lakatos claims that his characterization of science provides a method of objective assessment of the growth of scientific knowledge. Since Lakatos's methodology of scientific research programmes developed out of his extended version of sophisticated falsificationism, the definitions presented earlier in relation to that version of sophisticated falsificationism (for example, novel facts, progress, and so on) apply equally well to the methodology of scientific research programmes. Lakatos can therefore present an assessment of the growth of scientific knowledge in terms of progressive and degenerating problemshifts. A research programme is successful if it leads to progressive problemshifts and unsuccessful if it leads to degenerating problemshifts.[38] As we have seen, growth of scientific knowledge for Lakatos means empirical growth and a successful research programme has its progress measured by the rate at which its predicted novel facts are corroborated. Thus Lakatos concludes that a research programme will display an objective growth if each of its phases are accurately characterized as progressive theoretical problemshifts and its predicted novel facts are (from time to time) experimentally verified, that is, there is the occasional progressive empirical problemshift. As Lakatos says:

> we must require that each step of a research programme be consistently content-increasing: that each step constitute a *consistently progressive theoretical problemshift.* All we need in addition to this is that at least every now and then the increase in content should be seen to be retrospectively corroborated: the programme as a whole should also display an *intermittently progressive empirical shift.* We do not demand that each step produce *immediately* an *observed* new fact. Our term '*intermittently*' gives sufficient *rational* scope for dogmatic adherence to a programme in face of *prima facie* 'refutations'.[39]

Lakatos is careful to portray science as a rational exercise. (Recall the basic definition of 'rational' given on p. 52.) Both the autonomy of theoretical science from empirical science and the growth of scientific knowledge are provided with objective bases in Lakatos's methodology. We have already seen how objective growth is ascertained. The autonomy of theoretical science is guaranteed by the positive heuristic. It determines which problems are suitable for researchers to investigate and keeps them from becoming concerned about the anomalies present in the programme.[40] It is not the individual researcher who makes these determinations. Acceptance of a particular research programme makes the pursuit of only that research set out by the programme a rational act.

Contrary to the assertion of Lakatos, the presence of serious anomalies do count in the assessment of a research programme. This is borne out by the history of science. For example, inaccuracies in planetary positions and calendar discrepancies calculated from the Ptolemaic astronomical theory were contributing factors in the development of, and to the eventual acceptance of, the Copernican theory. In addition, Lakatos's claim that, in an empirically progressive research programme, theoretical science is autonomous from empirical science cannot be entirely correct. This is because theoreticians need empirical feedback on the testing of previous theoretical endeavours before further theoretical advances can be made.[41]

If a research programme is to be progressive then, under Lakatos's definition, it must make empirical progress, otherwise the programme is degenerating. Non-empirical advances (that is, theoretical ones), although being a prerequisite for empirical progress, do not of themselves provide progress for a research programme in terms of Lakatos's definition. Yet, in spite of Lakatos's claim for the autonomy of theoretical from empirical science, pure theoretical advances on their own fail to register on a research programme's progress 'scale', as one might expect they should.

The methodology of scientific research programmes offers its own solution to the problem of science–non-science demarcation.

Science is distinguished from other forms of knowledge by the criteria which specify what constitutes a *scientific* research programme. There are other knowledge pursuits that roughly conform to the structure of research programmes. In Lakatos's opinion, however, these programmes cannot be considered scientific. In order to qualify as scientific, Lakatos says that research programmes must satisfy two criteria:

- the theories within a research programme must be unified by a central theme; they must exhibit a strong continuity and these theories must possess heuristic (explanatory) power. The programme, through its positive heuristic, must indicate lines of future research.
- research programmes must be empirically progressive, that is, they must occasionally yield novel predictions which are in time experimentally verified.[42]

Lakatos cites the example of Marxism as satisfying the first criterion but not the second. Marxist theories make many predictions most of which would be expected or highly probable given the theory, but Lakatos claims that they also make small numbers of novel predictions, none of which has been corroborated. Under a Lakatosian analysis then, Marxist theories are not scientific.[43] The example of statistically based theories of social psychology is cited as a set of theories for which the second criterion holds but not the first:

> [statistically based social psychology] hits patched-up, unimaginative series of pedestrian 'empirical' adjustments which are so frequent, for instance, in modern social psychology. Such adjustments may, with the help of so-called 'statistical techniques', make some 'novel' predictions and may even conjure up some irrelevant grains of truth in them. But this theorizing has no unifying idea, no heuristic power, no continuity.[44]

DEGENERATING RESEARCH PROGRAMMES AND SCIENTIFIC REVOLUTIONS

In this section we will address the question of what happens when a research programme starts to degenerate. A research programme will degenerate if it is not empirically progressive—if its predicted novel facts do not receive experimental confirmation. We have seen that a successful research programme need only occasionally be empirically progressive—have intermittent empirically progressive problemshifts. Since this is all that is required for a research programme to be progressive, there may well be periods in the lifetime of a progressive research programme when its predicted novel facts have not received experimental verification. If such a period continues for long in comparison to (say) the previous time intervals when the programme has not been empirically progressive, then this longer period could be characterized as a degenerating phase in the research programme's life. Such a degenerating phase indicates that the positive heuristic of the programme is no longer offering sufficient guidance of a suitable type to the researcher. It might be said that in such circumstances the positive heuristic has 'run out of steam'! Degeneration is also marked by both a failure to predict novel facts and by attempts to account for anomalies in the programme—for example, by *ad hoc* explanations:

> The Newtonian programme led to novel facts; the Marxian lagged behind the facts and had been running fast to catch up with them . . . where theory lags behind the facts, we are dealing with miserable degenerating research programmes.[45]

Degeneration is not necessarily irreversible. A research programme may again be made progressive. This can be achieved by making changes to the positive heuristic. The positive heuristic of a programme is, in general, a fairly flexible set of guidelines and therefore readily open to alteration as the situation demands.[46] Note that no external assessment of a research programme is required to indicate whether degeneration is present in the programme. One yardstick by which degeneration is measured is the lack of empirical progressiveness over a time period sufficient for progress to occur. A degenerating phase of a research programme is really only seen as such in hindsight. At the time, it may not be clear whether the apparent degeneration is merely one of the intermittent periods between corroboration of novel facts, or whether the programme has gone into a degenerating phase. Only after there have been some changes made to the positive heuristic, which result in an empirically progressive problemshift, is it correct to claim that there has been a degenerating phase from which the programme has recovered.

We shall now introduce the analytical method of rationally reconstructing the history of science. We noted earlier Lakatos's claim that postdiction is as important as prediction, since empirical data have little significance unless explained by theory. When part of the history of science is reconstructed, the time ordering of empirical discoveries and/or theory proposal is altered to see what conclusions can be drawn. If the same evidence and theories are reconstructed in different time orders, a degenerating problemshift may be transformed into a progressive one or vice versa. Lakatos distinguishes between an external and internal history of science. The external history is secondary to the internal (that is, the rationally reconstructed) history. External history explains events only in so far as the internal history does not.

The modern revolution in geology was an episode in science containing many examples of what appeared to be degenerating problemshifts, but which on a rational reconstruction were found to be progressive. We shall consider one of these cases. The Cold War period of the mid-twentieth century saw, among other things, undersea mapping and other submarine research on a scale not previously imagined. This research was primarily funded by the United States Office of Naval Research. The prevailing view towards much of this research was that large amounts of empirical data about the ocean floor should be collected and analysed at a later time—data collection was made the priority.[47]

Two sets of data were very important in bringing about the modern revolution in geology. The first was the finding that the ocean floor was relatively young compared with continental landmasses. This was testified to by both the age of rocks which had been dredged up and seismic soundings which indicated sediments covering the ocean floor were much thinner than expected.[48] The relative youth of the world's ocean floors was well accepted by the end of the 1950s. The second set of data was ocean floor magnetic profiles collected off North America's west coast prior to 1962 by the Scripps Institution of Oceanography. These profiles showed that adjacent 'strips' of seafloor (each twenty to thirty kilometres wide) displayed an alternating pattern of magnetization. The rocks forming one such strip would be magnetized in a particular direction whilst the rocks in the two adjacent strips would be magnetized in the opposite direction. This was indeed a curious finding and had no theoretical explanation.[49]

In 1962 the American geophysicist Harry Hess proposed the 'Seafloor Spreading' version of continental drift. In this theory, Hess claimed that new seafloor is formed at undersea ridges. Material from deep inside the Earth 'wells-up' and pours out on both sides of a ridge to spread the existing ocean floor further away from the ridge.[50] This hypothesis had no direct empirical support at the time of its proposal. In 1963 Fred Vine and Drummond Matthews of the University of Cambridge explained the magnetic profiles collected by the Scripps Institution in the context of a seafloor which is created at mid-ocean ridges and spreads uniformly on either side of the ridge, creating a pattern of magnetic striping as it spreads.

Let's now retell (reconstruct) this story by changing the order in which these events occurred. Suppose that Hess had proposed his theory before the magnetic profiles were known. Hess's theory would constitute theoretical progress for the continental drift research programme. It would have predicted novel facts related to the spreading and creation of new seafloor. Further suppose that after the proposal of the seafloor spreading theory, the ocean floor was then found empirically to be quite new. This finding would corroborate the predictions of Hess's seafloor spreading and therefore would count as empirical progress (or an empirical problemshift) for the drift programme.

If we continue with this rational reconstruction, then the next piece of the reconstructed picture would be the proposal of the Vine–Matthews hypothesis with its prediction of the adjacent strips of seafloor oppositely magnetized. Such strips would have directions of magnetization first one way and then reversed if the Earth's magnetic field periodically underwent a reversal, as was indicated by other studies unrelated to the seafloor. The Vine–Matthews theory would then constitute a further theoretical problemshift for the continental drift research programme. Lastly, the discovery of alternating strips on the seafloor would be a dramatic confirmation of the Vine–Matthews theory and would therefore continue the empirical progressiveness of the programme. Without the theories of seafloor spreading, the magnetic data would have been nothing more than an unexplained curiosity and without the recognition of the significance of the magnetic data, continental drift in the form of seafloor spreading would have appeared to be a degenerating research programme.[51]

Lakatos claims that theoretical pluralism is the normal situation in science; that is, there usually exists more than one research programme in a given discipline at any time. If this is the case, and historical evidence does tend to bear it out, then the question arises as to how a research programme can be eliminated from competition. (This is not a reiteration of the sophisticated falsificationist criterion of 'falsification' on the emergence of a better theory, for that refers to individual theories contained in a single research programme.) It is not sufficient to eliminate a programme solely on the basis that it presently appears not to be making empirical progress. A programme may only be in a degenerating phase from which it will recover, or a reconstruction of the history may show the programme's problemshifts were, in fact, progressive ones.[52]

If scientific research is conducted in an objective manner—if its methods, procedures and decisions are not idiosyncratic or are not dictated merely by social or political circumstances—then it should be possible to specify rational reasons for abandonment of a research programme. As part of his methodology, Lakatos puts forward a criterion for elimination which he claims is objective and he also claims that a decision to eliminate a research programme made on the basis of this criterion is fully rational. He summarizes this as follows:

> *Can there be any objective* (as opposed to socio-psychological) *reason to reject a programme, that is, to eliminate its hard core and its programme for constructing protective belts?* Our answer, in outline, is

> that such an objective reason is provided by a rival research programme which [1] explains the previous success of its rival and [2] supersedes it by a further display of *heuristic power*.[53]

These two considerations need some amplification. The first is analogous to the situation of the 'falsification' of one theory in a series of consecutive theories within a research programme by the next one in the series. Any research programme which has only just emerged must account for the same 'facts' as those predicted by the existing research programme and do so by its own (new) explanations. New explanations are to be expected as the new research programme will have different hard-core assumptions and a separate positive heuristic than did the previous programme. A new research programme can be expected to explain the existing 'facts' in its own terms straight away, but it is too much to expect it to immediately produce novel facts.[54]

A good example is the Copernican heliocentric astronomical theory. Copernican theory gave only slightly more accurate predictions of planetary positions than did Ptolemaic theory and did so by its own mechanisms; for example, the Copernican system had a central Sun and a moving Earth. The 'old facts' of Ptolemaic theory were immediately explicable in a 'novel way' by the theory of Copernicus. It took a little time before novel facts were being predicted by Copernican theory, such as the complete set of the phases of the planet Venus.[55]

In general a new research programme will take a considerable amount of time to become empirically progressive. If there is a long time-lag between the advent of the new programme and when it first predicts novel facts (that is, the time to become theoretically progressive) then it stands to reason that it must take an even longer time period before such facts can be experimentally corroborated. In some extreme cases this time period could be centuries. A case in point was the discovery of the annual stellar parallax predicted by Copernican theory. If the Earth was really revolving about the Sun as Copernicus had postulated, then there should have been an observable difference in the angle between two stars when this angle was measured at six-monthly intervals. The confirmation of this prediction had to wait almost three hundred years until technical advances allowed the extremely small difference in the two angles to be measured.

A new research programme must be given sufficient time to show its heuristic power—its ability to generate new explanations and novel facts. In addition, there must be an allowed time (sometimes a very long time) for the novel facts to be confirmed. Only by the exercise of such methodological tolerance can a new research programme supersede an existing one; that is, overtake its rival (or rivals) by exhibiting greater heuristic power and empirical progressiveness. During the stages of a new programme's life leading up to its overtaking of a rival, commitment to the new research programme may be judged as being rational or not rational by means of a rational reconstruction evaluation. A decision to accept or not to abandon a new research programme is rational if in the absence of its rival(s), the new programme is correctly characterized as being progressive.[56]

Given that a new research programme is allowed time to become empirically progressive, this does not automatically mean it will supersede a rival and thereby eliminate it from competition. The competition between rival research programmes may continue for centuries before there is a clear winner. The competition between Copernican and Ptolemaic research programmes is one such example. The reason for this is that there is no such thing as a 'crucial experiment' which can be used to make a final decision between rival programmes. Instead there are usually many experiments, the outcomes of which may favour either the old programme or the new. Thus there can develop a situation in which the leading place in the competition swings to and fro between the established research programme and a challenger. How can this happen? Adherents to each programme will continue to produce better and better versions of their respective programmes; that is, new theories within their own programme with each individual theory having increased content over its predecessor. Verification of some of this increased content (or corroborated novel facts) will, for the research programme concerned, constitute a progressive problemshift. If the rival programme can be made to perform similarly, then it becomes progressive and surges ahead. Lakatos writes:

> When two research programmes compete . . . the *n*-th version of the first will be blatantly, dramatically inconsistent with the *m*-th version of the second. An experiment is repeatedly performed, and as a result, the first is defeated in *this battle*, while the second wins. But *the war* is not over: any research programme is allowed a few such defeats. All its needs for a comeback is to produce an *n*+1-th (or *n*+*k*-th) content-increasing version and a verification of some of its novel content. If such a comeback, after sustained effort, is not forthcoming, the war is lost . . .[57]

(Here *m* and *k* represent integers that are greater than one.)

It is now possible to identify the means by which the choice between rival research programmes is made and how research programmes are eliminated. At some stage in the competition between rival programmes (usually after the competition has been proceeding for some time) one programme remains progressive whilst the others only degenerate. A scientist entering a field of research within which there exists more than one research programme has to make a choice about which programme to join. The choice in this case should be for the programme that is displaying empirical progressiveness. So too should be the choice for a researcher already working in a research programme. The choice either to abandon one's present research programme or transfer to another should be determined by the empirical progress of the competing research programmes.

When a research programme overtakes its rivals and continues to make progress, this results in the effective elimination (in the long term) of the rivals from competition. This superseding of one research programme by another defines a 'scientific revolution' within Lakatos's theory. A scientific revolution has occurred when a progressive research programme supersedes its degenerating rivals. The progressive programme will take with it the majority of researchers, simply because it is making progress while the others are not. Lakatos rhetorically poses and then answers the question of what is a scientific revolution:

> Now, how do scientific revolutions come about? If we have two rival research programmes, and one is progressing while the other is degenerating, scientists tend to join the progressive programme. This is the rationale of scientific revolutions . . . Kuhn is wrong in thinking

> that scientific revolutions are sudden, irrational changes in vision. The history of science refutes both Popper and Kuhn: on close inspection both Popperian crucial experiments and Kuhnian revolutions turn out to be myths: what normally happens is that progressive research programmes replace degenerating ones.[58]

The Einsteinian Revolution is, perhaps, the prime example of one research programme surpassing another and causing a scientific revolution. At the beginning of the twentieth century, physics as a discipline seemed to be uncontroversial. The accepted framework of Newtonianism was highly successful and there appeared to be no major unsolved problems outstanding. It looked as if the physical universe could (and eventually would) be explained within the confines of Newtonian mechanics and gravitational theory.[59] We know that Albert Einstein was to change this. In his 'General Theory of Relativity', Einstein explained why the equations of motion for all the planets (except for Mercury) were derivable from Newton's theory of universal gravitation and the analysis of planetary perturbations. In the case of Mercury there was a slight, observable departure in its motion from that predicted by Newtonian theory. (This phenomenon is the well-known precession of perihelion, where the position of Mercury's closest point to the Sun changes with every orbit.) General relativity not only predicted this so-called 'anomalous' motion of Mercury, it also showed why the Newtonian solution did not predict the observed value for perihelion precession. Einstein's theory not only explained Newtonian results in a 'novel' way, it also predicted novel facts: for example, the bending of light rays that pass close to the Sun. This and several other important predictions were quickly corroborated within a few years. It can therefore be claimed that Einstein's general relativity superseded the Newtonian research programme by its display of heuristic power and its rapid empirical progressiveness.

RATIONALITY AND THE AIM OF SCIENCE IN LAKATOS'S THEORY

What is the aim of science in Lakatos's theory? Lakatos acknowledges that there cannot be an empirical basis for science of which we can be certain and that there are no such things as non-theory-laden facts. Clearly then, he cannot be advocating that his methodology *necessarily* yields truth as an end product. If the aim of scientific research is not truth about the external world (and not because truth is undesired but because it is inaccessible) then what is the aim of science? Lakatos rejects Kuhn's idea that the aim of science is merely to solve 'puzzles'. The value of solutions to Kuhnian 'puzzles' is subjective, according to Lakatos, since they are acceptable only if consistent with 'the collective wisdom of the [current scientific] community'. Lakatos, however, argued that scientific knowledge has an objective value. In this respect Lakatos agrees with Popper.[60]

Popper says that one aim of science is to obtain 'Objective Knowledge', but what exactly is objective knowledge and how does it differ from truth? Popper's response to this question would be to talk about the existence of three 'worlds' (for want of a better name). The 'first world' (or World 1) is the world of physical objects or physical states—those things and conditions which are contained in our physical universe. The 'second world' (or World 2) is the world of states of consciousness (or of mental states)—all thoughts and ideas conceived in the minds of people. The 'third world' (or World 3) is the world of objective contents of thought—logically consistent theoretical structures or frameworks which exist regardless of whether they have been conceived by anyone or not. Scientific knowledge (which is the end result of scientific endeavour) belongs, according to Popper, to World 3 and is totally independent of the individual's belief or claim to know.[61]

Popper offers, as an example of objective knowledge, a series of books containing tables of logarithms produced by a computer. He imagines that these books are sent to libraries but are not read for years and even then many of the actual logarithmic figures are never used by anyone. Yet each logarithm represents a mathematical theorem and as such is a form of objective knowledge. Popper explains this notion:

> The example of these books of logarithms may seem far fetched. But it is not. I should say that almost every book is like this: it contains objective knowledge, true or false, useful or useless; . . . [what makes]

> black spots on white paper into a book, or an instance of knowledge in the objective sense . . . is its possibility or potentiality of being understood, its dispositional character of being understood or interpreted, or misunderstood or misinterpreted, which makes a thing a book. And this potentiality or disposition may exist without ever being actualised or realised.[62]

A number of points in this quotation need clarification. Popper says that the objectivity of scientific knowledge is guaranteed by its dispositional character. This is an intrinsic attribute (or disposition) which objective knowledge possesses to be known and interpreted by conscious beings. The best example of objective knowledge is mathematical knowledge. There is a strong sense, according to Popper, Lakatos and other philosophers (notably Frege), in which mathematical laws and relations exist independently of anyone knowing or writing them down. Expression of many known and unknown mathematical relations appears all around us, reflected in the macroscopic order of the cosmos. One specific example is the theorem of Pythagoras: for any right-angled triangle, the square of the length of the hypotenuse is equal to the sum of the squared lengths of the other two sides. Popper and Lakatos would claim that Pythagorean Theorem is (timelessly) a member of World 3. When somebody first thought of the theorem (it was known prior to Pythagoras) it became a member of World 2. The study of relations between objects in our physical universe (in World 1) leads many to the conclusion that there are good reasons to believe that the theorem of Pythagoras is also part of our universe (World 1).

Similarly, it is claimed by those of this opinion that scientific theories also have an objective existence as the contents of World 3. The *content* of scientific theories is independent of individuals, their beliefs and social environment. This objective character is taken as part reason why scientists living in different times or places can independently postulate the same theories. It should be noted that although scientific knowledge in Lakatos's theory has an objective status, it need not be true. Objective knowledge can be either true or false. The overall truth or falsity of a scientific theory does not affect its objectivity, that is, its being a member of World 3. The scientific laws which govern our universe (whether known to us or not) could have been different from what they are. Since World 3 contains all scientific theories (whether actualized in our physical universe or not and where known or unknown) there must be scientific theories which are false—which are not true in our universe (World 1). There are also theories which Lakatos would want to claim are neither true nor objective. In this collection of theories, he would group such theories as Marxism and Freudian psychology.[63]

The arguments presented by Popper and Lakatos for the objectivity of scientific knowledge do not establish objectivity in a metaphysical sense. At most their arguments only establish the inter-subjectivity of scientific knowledge, that is, the knowledge need not be objective but may be known by several individuals quite independently of each other. For example, contemporaneous scientists not in communication may well develop the same theory autonomously. This is not really a surprising situation if one acknowledges that both scientists seek to explain the same phenomenon and both use similar experimental techniques and apparatus.[64] This, however, implies little about the theory they develop having a particular metaphysical status (for example, membership of Popper's World 3).[65]

We return now to the question of the rationality of science according to Lakatos. He describes science as a fully rational exercise yet the methodology of scientific research programmes does not provide the means to make an immediate rational choice between rival research programmes. The lag between emergence of a new research programme and its becoming empirically progressive shows that this choice may be delayed until there is a clear winner in the competition. (Sometimes this is expressed as there being an end to the idea of 'instant rationality'.)[66] These considerations also indicate that a scientific revolution is recognizable only with hindsight, long after one research programme has superseded another. Neither can it be said to be necessarily irrational for a researcher to continue working within a degenerating research programme.

An assessment of whether an individual researcher's commitment to a degenerating programme is rational or not will depend on a number of factors. For instance, if the research programme has only just been superseded by a new rival then the researcher may have good reason to think that the programme can (with much effort) be made progressive once more. If, however, many years have been devoted to a degenerating research programme and it has failed to recover, then the decision to remain with the programme for so long might be judged not to be fully rational. As before, such verdicts can really only be given with hindsight.

Lakatos would also deny that research programmes are incommensurable with one another as he claimed that their contents could be compared and their relative progress objectively ascertained.

The above conclusion regarding the rationality of choice between research programmes allows for a further judgement about the status of scientific knowledge in Lakatos's theory. Scientific knowledge in this context is better (or superior) to other forms of knowledge. Scientific knowledge is superior due to its objective content, objective growth and its rational generation. The rational norms of science are its heuristic principles and objective criterion for the elimination of rivals.

Lakatos's methodology does not, however, offer the scientist explicit and definitive criteria for choosing between rival research programmes. When a scientist makes such a choice, it can only be shown to be 'correct' with hindsight. This has been a major criticism levied at Lakatos, for although his theory provides a means of assessing research programmes (determining whether they are progressive or not), his accompanying methodology does not provide a means of making an unambiguous choice between programmes. The popular expression of this by philosophers is that Lakatos's methodology does not give advice to scientists for choosing between competing research programmes.[67] (Recall that there is always a chance that a superseded research programme may, with sufficient effort, become progressive again and thereby justify its adherents' faith in it.) Paul Feyerabend views the failure of the methodology of scientific research programmes to supply strict selection rules as a failure of rationality in Lakatos's theory: 'Reason as defined by Lakatos does not *directly* guide the actions of the scientist. Given this reason and nothing else, "anything goes"'.[68]

Lakatos has attempted to defuse such criticisms by further elaboration, although this response seems less than sufficient:

> The methodology of research programmes was criticised both by Feyerabend and by Kuhn ... [Musgrave also] demanded that I specify, for instance, at what point dogmatic adherence to a programme ought to be explained 'externally' rather than 'internally'.
>
> Let me try to explain why such objections are beside the point. One may rationally stick to a degenerating programme until it is overtaken by a rival *and even after*. What one must *not* do is to deny its poor public record. Both Feyerabend and Kuhn conflate *methodological* appraisal of a programme with firm *heuristic* advice about what to do. It is perfectly rational to play a risky game: what is irrational is to deceive oneself about the risk.[69]

4
Laudan's Theory of Evolving Research Traditions

Laurens (Larry) Laudan had been a graduate student studying astrophysics before he became interested in formal logic and the philosophy of logical inference. After working on problems of logic for a period, he realized that his strongest interests in philosophy lay in the examination of scientific rationality and the workings of science. The majority of his writings since that time have been devoted to the history and philosophy of science. Larry Laudan has taught and researched at various institutions including the University of Pittsburgh, the University of London, Virginia Polytechnic and State University, the University of Hawaii and the University of Melbourne. He was a co-founder of the British journal *Studies in History and Philosophy of Science.*

Laudan arrived at his theory of evolving 'Research Traditions' after much deliberation on the merits and shortcomings of earlier theories of science, especially those of T. S. Kuhn and Imre Lakatos. Laudan's theory utilizes what he takes to be the best aspects of the theories of Kuhn and Lakatos. He readily acknowledges his debt to these (and other) philosophers who have 'paved the way'. In his theory Laudan was particularly concerned with problem solving, with progress and the relations these have to the question of the rationality of science. This was detailed in his book *Progress and Its Problems.* The relationship between what is a rational choice and what counts as progressive in science is a keystone in his theory. A full account of this relationship has been a topic sorely neglected, in Laudan's opinion, by past philosophers of science and an explication of its implications is one task to which he addresses himself:

> Insofar as rationality and progressiveness have been linked at all, the former has taken priority over the latter—to such a degree that most

> writers see progress as *nothing more than* the temporal projection of a series of individual rational choices . . . I am deeply troubled by the unanimity with which philosophers have made progress *parasitic* upon rationality . . . The two concepts are doubtless related, but not necessarily in the manner usually supposed.[1]

EMPIRICAL PROBLEMS

Laudan's analysis of science centres on problem solving and its implications for the rationality of the scientific enterprise. It is not the case that Laudan thinks science is only a problem-solving activity, but when science is viewed as such he believes that much more is revealed than has previously been shown. This is why he places such an emphasis on problems and their solution. Problem solving is then the primary aim of scientific research according to Laudan. By concentrating on problem solving as the aim of science Laudan presents a model which, although more successful in accounting for rational choice in science than its predecessors, does so by mostly ignoring other legitimate aims of scientific research. Laudan's model tends to reflect one dimension of what is a multifaceted enterprise.

Scientists develop theories in response to existing problems and it is hoped that these theories provide solutions to those problems. Laudan argues that the assessment of scientific theories should be done on the basis of whether or not they offer sufficiently good solutions to problems posed.[2] What, however, makes one problem the subject of a strong research effort to find a solution rather than some other problem? What, for that matter, defines a 'something' to be a problem requiring a solution? Since problems and problem solving are central to Laudan's theory, he goes to some trouble to define what problems are and what roles they play in the scientific context. He initially separates problems into two broad categories: *empirical* problems and *conceptual* problems. We shall deal first with the nature, types and role of empirical problems.

Earlier we saw that Kuhn had defined an empirical problem to be some difficulty in matching theory and experimental results, but what counted as theory and what as fact was paradigm dependent. This is, in large part, due to the theory-ladenness of observation. Laudan accepts that all observations are theory-laden but none the less asserts that scientists (and people in general) accept many so-called 'facts' as being truths about the world. In the realm of an accepted theory it is taken for granted that 'facts' are indeed true features of reality, rather than some theoretical construct dictated by the acceptance of a particular theory. This is also the case for problems which scientists seek solutions. What counts as an empirical problem is partly dependent on what theories are currently held.[3] What then, does Laudan define as empirical problems? He offers a fairly wide definition—an empirical problem is anything in the natural world (or accepted as existing in the natural world) that strikes us as needing explanation.[4]

This (generous) definition needs some clarification. Firstly, theory-ladenness is included by Laudan stating that empirical problems are part of what we take to be in the natural world (that is, those things that we perceive to be real due to our biased presuppositions about nature). Secondly, in order to constitute an empirical problem those features which we take to exist must also be seen as requiring some explanation. Finally, 'facts' (even though they are theory-laden) are pragmatically distinguished in each system of theoretical beliefs from the constituent theories of that system. (This is the case regardless of whether these facts actually relate to anything real or not.)

However, if a fact is considered to require some explanation then, in Laudan's view, there has to be a need for (or value placed on) gaining such an explanation. Not all facts are considered empirical problems; if there seems no need to explain a fact then it is, by definition, not a problem:

> To regard something as an empirical problem, we must feel that *there is a premium on solving it.* At any given moment in the history of science, many things will be well-known phenomena, but will not be felt to be in need of explanation or clarification. It was known since the earliest times, for instance, that most trees have green leaves. But that "fact" only became an "empirical problem" when someone decided it was sufficiently interesting and important to deserve explanation.[5]

Thus, what counts as an empirical problem at one time may not be seen as a problem at a later time, or vice versa.[6] What perhaps brings out the distinction between facts and empirical problems is that facts cannot change in the way which empirical problems can. In Laudan's above example, one can say that a leaf is green whether or not we see that as something to be explained.

Laudan divides empirical problems into three classes: unsolved problems, solved problems and anomalous problems (anomalies). We shall deal with each in turn. Unsolved problems have some curious characteristics. Their status is somewhat ambiguous and for the purposes of assessing the merits of rival theories, unsolved problems are considered irrelevant by Laudan. A given unsolved problem may turn out not to be a problem at all. A conclusion to this effect might be reached by improvements in instrumentation, or by a change of theory. How then, does one know that an unsolved problem is a genuine problem? Laudan's answer is that we don't know until the problem is solved by a theory. Until that happens unsolved problems are really only potential problems.[7] In regard to the appraisal of rival theories, Laudan argues that if a theory has not solved a particular problem (and no other theory has solved the same problem), then this cannot be held against the theory in question. Why? Because until the problem is solved there is no way to know if it can be solved in the relevant domain of the theory or not.[8]

The question now arises: what constitutes a solved empirical problem? Laudan's general response is that empirical problems are solved when scientists working in the relevant discipline believe that they have gained an understanding of why the problem situation exists in that form. Empirical problems are solved by theories. Since Laudan distinguishes facts from empirical problems, solving such problems is not equivalent to merely 'explaining the facts'. Now Laudan is under no misapprehension about the status of solved problems. Any solved problem is solved within the theoretical boundaries of some theory, be that theory true or false. Therefore the truth or falsity of a theory is not relevant for the purposes of problem solving.[9]

Laudan maintains that a solution to a problem does not have to be exact. For instance, a quantitative (that is, numerical) expression of the solution does not have to agree in every respect with empirical findings. The scientist accepts this sort of situation without question, for example, when experimental results do not precisely match with theoretical predictions but are none the less within accepted bounds of error. The facts obtained via experiment are not literally explained since the theoretical prediction is not exactly the same as the empirical result. Laudan's claim is drawn from actual scientific practice: he cites the examples of Newton's explanation of planetary motion, Einstein's prediction of the bending of light rays, inexact chemical bonding theory predictions and heat transfer in steam engines.[10] If the propositions of a theory (logically) entail a statement of a problem (even if the statement is not an exact expression of the problem) then it may be said that this theory solves the problem.

The irrelevance of the truth of a theory may initially sound quite strange but theories do not have to be true to give correct predictions. This is an important point to note. The level of empirical adequacy (how close the agreement is between theory and experiment) is not a measure of the truth-content of a theory. False theories can work just as well as true ones. Indeed, the history of science is full of discarded theories which during their lifetimes were accepted as true because they worked so well.[11] An example of a 'refuted' theory that was well accepted is Newtonian 'Corpuscular Optics'. In this theory light was assumed to be made up of 'swarms of hard particles' which obeyed the laws of Newtonian mechanics. The theory explained the processes of reflection, refraction and dispersion quite well. Corpuscular optics was, therefore, successful in solving problems related to these processes (at least for a time). The wave theory of light also solved the same problems but these two theories cannot both be true. If light is a hard particle which is localizable at a particular position it cannot also be spread out in space in the manner of a wave. Yet both theories offered solutions to the same problems.

The differences between the concepts of a solution to a problem and an explanation can be brought out more clearly by reference to the above example. Both wave and particle theories of light offer solutions to empirical problems. Both theories offer explanations for observed phenomena (although different and mutually inconsistent ones). However, there is a sense in which the solutions provided by one theory are better than the solutions provided by the other (in this case the wave theory is better). 'Better' in this context means a closer agreement between prediction and experiment. In Laudan's view no such comparison is possible in terms of the explanations offered by theories:

> the notion of solution is highly relative and comparative in a way that the notion of explanation is not. We can have two different theories which solve the same problem, and yet say one is a better solution, (i.e., a closer approximation) than the other.[12]

Laudan's claim that the adequacy of solutions to problems changes from time to time seems fairly uncontroversial. Scientific standards change over time and do generally tend towards greater accuracy, greater precision, more rigour and, as a consequence, more complexity. No one would seriously suppose that accepted solutions to empirical problems at the time of Aristotle would be acceptable to today's physicists. This facet of science is seen by Laudan as a strong and important feature.[13]

We turn now to anomalous problems (anomalies). Laudan proposes a new definition for these which differs from the meanings ascribed by Kuhn and by Lakatos. Laudan defines anomalous problems as the empirical problems that have not been solved by a theory but which have been solved by one or more competing theories.[14] In Laudan's view an anomalous problem (also called a 'Laudanian anomaly') need not be inconsistent with a theory for which it is an anomaly. An unsolved problem may be perfectly consistent with the basic underpinning of a given theory and yet be an anomaly for that theory simply because it has not been solved by the theory, but by a rival theory. One example is the problem of the apparent 'fit' of the coastlines of Africa and South America for the rival theories of geology as they existed in the early 1900s. Wegener's theory of continental drift had a perfectly natural solution to this problem: the continents were once joined, they broke up and drifted apart.[15] This solution created an anomaly for the theory of permanentism which held that the continents and ocean basins were permanent features of the Earth's surface. If it is the case that the oceans and continents are permanent, how does one then account for the congruence of the coastlines? This is not a problem which is necessarily inconsistent with continents and ocean basins being permanent. However, some alternative explanation which does not involve laterally moving continental masses would be needed to convert this anomaly into a solved problem for permanentism.

It is worth digressing for a moment to compare the various senses of anomaly canvassed so far. The Kuhnian anomaly is an unsolved problem which has (somehow) become so important that it is viewed as a counter-instance to a theory. Although Kuhn is not exact about how this change in perspective comes about, anomalies play a central role in his scheme of science. The emergence of one or more Kuhnian anomalies precipitates a crisis state in a scientific discipline which results in a loosening of the rules that govern normal science. In Lakatos's theory, an anomaly can merely be some unexpected event that has appeared, a difficulty in matching predicted to experimental results, or an inconsistency in the theory itself. Lakatos claims that anomalies can be ignored; that they are not important as long as empirical progress is being made. Laudan, on the other hand, is quite specific in his definition of anomaly. This specificity has the distinct advantages of not only removing any ambiguity in the notion of anomaly but also providing a method whereby the transition from an unsolved problem to an anomaly is clear cut. Laudan holds that, if an anomaly exists for a particular theory, this does not necessarily require researchers to give up that theory, but the anomaly does bring the theory into serious question.[16]

A fact becomes an empirical problem, for Laudan, only when there is a value placed on explaining it. Further to this, he claims that not all empirical problems are of equal importance: different empirical problems can be weighted differently. Laudan says that there are at least four methods by which the weight—importance—of an empirical problem is increased and three by which the problem weighting is decreased. When a new and distinct scientific discipline appears, there will initially be no solved empirical problems, only potential ones, and these would be of equal importance. Once a theory in the field begins to produce solutions, the weights attached to different problems change.

There are four methods of increasing a problem's importance ('problem inflation'). The first method is 'Inflation by Solution' (unsolved to solved status)—if a theory solves a problem, then that problem acquires a greater significance. Not only does the solution show that the problem was a legitimate one for the relevant field but, if there are rival theories in the field, then its solution makes the problem an anomaly for these rival theories.[17]

The second method is 'Inflation by Anomaly Solution'—the significance of a problem is raised by its solution when it had previously constituted an anomaly for the theory which solves it. Why this should be a different situation with a different weighting to the solving of an (otherwise plain) unsolved problem is explained by Laudan:

> Unlike the solution of some new problem, the conversion of anomalies into problem-solving successes does double service: it not only exhibits the problem solving capacities of a theory (which the solution of any problem will do) but it simultaneously eliminates one of the major cognitive liabilities confronting the theory.[18]

The third method is 'Inflation by Archetypal Construction'—some empirical problems may be identified within a given theory as being fundamental to that particular theory. These problems are denoted as 'archetypes', meaning that they form a basic set of problems from which other problems in the field can be reduced.[19] The method of selection in Darwin's theory of evolution was, for example, a primary (archetypal) problem for the theory.[20] Darwin postulated that surrounding environmental conditions dictate which species live and which perish; those species best adapted to their surroundings will naturally stand the best chance of raising viable progeny and thereby continuing the existence of their species. This is fundamental to his theory of evolution by natural selection.

The final method is called 'Weighting by Generality'. Suppose that the solution to one problem is also the solution to a second problem, but that the converse is not the case. Then the first problem is said to be more general than the second and therefore the first problem carries a greater weight.[21]

There are also at least three methods for reducing the importance of an empirical problem ('problem deflation'). Firstly, there is 'Deflation by Dissolution'—the nature of empirical problems is such that what constitutes a problem today may not be seen as a problem in the future. Thus the problem and/or its importance may disappear altogether if reasons (empirical or otherwise) are found for believing it is no longer a problem.

Secondly, there is 'Deflation by Domain Modification'—if a problem is seen to become part of another scientific discipline (particularly if theories in that discipline successfully solve it) then the importance of the problem in the original field is lowered.

Lastly, there is 'Deflation by Archetype Modification'—in this case empirical problems that were considered archetypes in any theory which has been discarded will tend to be devalued by the process.[22]

It is for anomalies that the attached weighting is of particular importance since this is a crucial factor involved in any decision to reject a theory. Laudan was especially critical of both Kuhn and Lakatos on their superficial treatments of the nature and role of anomalies. Anomalies are the most important empirical problems and potentially the most damaging for the theories in which they occur. Anomalies, like other problems, can be given a weighting according to their importance to a theory. Some anomalies merely have nuisance value while others may devastate a theory.[23]

Laudan claims that the importance of a particular anomaly can vary greatly depending on a number of circumstances. The process of rating the importance of anomalies cannot be performed in a vacuum. Any assessment has to be performed in the context of the scientific discipline in which it occurs. In particular the assessment will be different in circumstances where several rival theories have solved the problem, since this gives strong reason to reject the theory for which the problem is an anomaly. Thus, Laudan argues that the importance of an anomaly is to be judged on the basis of how much it would tend to make researchers abandon the theory for a rival. Laudan labels this 'the degree of epistemic threat', one which has arisen due to the acquisition of some piece of knowledge—in this case empirical evidence that conflicts with theoretical predictions. Two other considerations rate a mention by Laudan as factors which affect the weighting of anomalies: first, the size of the discrepancy between the experimental result and predicted value; second, the time and effort put into trying to solve an anomaly.[24] This second point has a parallel in Kuhn's theory. Kuhn claims that one possible cause of the change in the perspective of a problem from unsolved to anomaly status is how long a problem has resisted concerted attempts at solution.

CONCEPTUAL PROBLEMS

Conceptual problems form the second broad category of problems. These are problems of a non-empirical nature which have a major role to play in the evaluation of rival theories. Laudan criticizes earlier philosophers of science for paying little or no attention to conceptual problems. Those that did notice them tended, in Laudan's estimation, to draw incorrect conclusions from their presence:

> Some scholars (such as Kuhn) have gone so far as to make the absence of such nonempirical factors a token of the "maturity" of any specific science. Rather than seeking to learn something about the complex nature of scientific rationality from such cases, philosophers (with regret) and sociologists (with delight) have generally taken them as tokens of the irrationality of science as actually practiced.[25]

Laudan defines two types of conceptual problems, 'internal' and 'external'. Internal conceptual problems arise from logical inconsistencies in the theory or through either ambiguities in its concepts or circularities in its schema. The first of these—inconsistency—is the most serious. No theory that contains logical inconsistencies (for example, predicts both that an event will occur and that it will not occur at the same time) can be accepted. However, as Laudan notes, inconsistency does not prevent researchers working on such a theory in order to resolve the inconsistency.[26] The other forms of internal conceptual problem—ambiguities and circularities—are no barrier to acceptance of a theory. Ambiguities arise when the theory terms are not sufficiently specific and therefore lend themselves to wide interpretation. Circularities occur when part of a theory's schema ends up deducing the original premises with which it began. The theory of quantum mechanics, for example, contains many internal conceptual ambiguities but this in no way prevents it from being a highly successful theory at the empirical level. The ambiguities in quantum mechanics are evidenced by the number of different interpretations of the theory: for example, statistical, nondeterministic, causal, non-causal, realist, instrumentalist, many 'worlds', propensity, local hidden-variable, non-local hidden-variable, and so on.

External conceptual problems are defined by Laudan as follows:

> External conceptual problems are generated by a theory, *T*, when *T* is in conflict with another theory or doctrine which the proponents of *T* believe to be rationally well founded. It is the existence of this "tension" which constitutes a conceptual problem.[27]

By 'conflict' Laudan means one of the following three situations:

- two theories are logically inconsistent with each other (although each individual theory is itself logically consistent);
- two theories are logically compatible but taken together they seem to infer rather implausible situations;
- a theory is compatible with another and should add credence to it, yet fails to do so.[28]

Laudan is very fond of using the adjective 'cognitive', especially when he is discussing conceptual problems. 'Cognition' literally means 'the act or faculty of knowing'. Science is a cognitive (or knowledge-producing) activity. Laudan suggests that there are five cognitive relationships that can exist between any two theories T_1 and T_2 (or parts thereof):

(1) entailment—T_1 logically entails T_2;
(2) reinforcement—T_1 gives a rationale for T_2;
(3) compatibility—T_1 does not entail T_2;
(4) implausibility—T_1 entails that T_2 is improbable;
(5) inconsistency—T_1 entails the negation of T_2.[29]

Situations that involve any of the relationships 2–5 can result in conceptual problems. Such situations pose, in Laudan's words, a 'cognitive threat' to the theories concerned, with the magnitude of the threat increasing from relationship 2 to 5.[30] Why should these be considered threatening to the relevant theories? The existence of conceptual problems in a theory provides a non-empirical criterion for casting doubt on whether a theory should or should not be accepted. The more serious the conceptual problems are, the more a theory is placed in doubt. This is the sense of 'cognitive threat', as will become clearer when we discuss the possible sources of external conceptual problems.

Laudan claims that there are at least three distinct classes of situation—intra-scientific, normative and world-view—in which external conceptual problem emerge.[31] We shall examine each of these three classes.

Intra-scientific

This is the situation when two scientific theories from different disciplines are inconsistent with each other. Such an inconsistency implies that at least one of the theories should be rejected.[32] When Wegener put forward his version of continental drift theory, he postulated that the continents had moved to their present global positions by 'ploughing through' the more dense seafloor. The question was then raised: what immense force powers such a move? Wegener suggested some possibilities but these were shown to be vastly inadequate.[33] In the absence of a suitable mechanism to push the continents, Wegener's theory of continental drift was in serious conflict with Newtonian physics. It should be noted that, in principle, this conflict is a problem for both Wegener's drift theory and any hypothesis of a physical driving mechanism for the continents. This conflict was used as a strong argument for the rejection of Wegener's theory by most English-speaking geologists at that time.

Another example will further illustrate implications of intra-scientific external conceptual problems. The Danish physicist Niels Bohr advanced his model of the atom in 1913. In this model it was postulated that the electrons surrounding the atomic nucleus had fixed, stable orbitals, or closed paths, in which the electrons would not radiate electromagnetic waves. This stipulation of Bohr's was contrary to the predictions of classical electrodynamics which state that all charged particles moving in a curved path (or otherwise accelerating) must radiate electromagnetic waves. Bohr had no theoretical reason to explain why these atomic electrons did not radiate; his response was a pragmatic one. He said that if an electron in a stable atomic orbital did radiate electromagnetic waves, this situation would lead to a continuous loss of energy from the electron and the atom would consequently collapse in a very, very short time interval. This does not happen. Bohr therefore concluded that atomic electrons in stable orbitals do not radiate electromagnetic waves. Despite Bohr's rhetoric, his theory remained in conflict with the well-established theory of classical electrodynamics.

As with the case of Wegener's theory, such conflict is two-way. It is a problem for both Bohr's atomic theory and for classical electrodynamics. In this case though, the external conceptual problem did not immediately result in rejection of either theory. Instead the conflict led to the development of two new theories of greater sophistication, quantum mechanics and quantum electrodynamics, which together solved the original problem of why atomic electrons do not radiate electromagnetic waves when in stable orbitals. The development of new theories from older ones is an alternative to outright rejection.

Normative

A second class of external conceptual problems occurs when a researcher violates the currently accepted norms of scientific practice when producing or developing a theory. In other words, if a scientist opts to use a method different from that which his or her peers hold to be correct (or at least appropriate) then this constitutes grounds for either rejection of the resulting theory, or perhaps, for the modification of the established methodological criteria. Laudan strongly suggests that the importance of such normative difficulties for a theory is not to be underestimated as these can become sources of major external conceptual problems:

> the fate of most of the important scientific theories in the past have been closely bound up with the *methodological* appraisals of these theories; *methodological well-foundedness* has been constitutive of . . . the most important appraisals of theories.
>
> It is precisely for this reason that perceived methodological weaknesses have constituted serious, and often acute, conceptual problems for any theory exhibiting them.[34]

A major criticism levied at Wegener's theory of continental drift was the methodology by which the theory was developed. The accepted methodology at that time for most British and American geologists was first to gather large amounts of empirical data and only then to arrive at a theory by the process of inductive generalization. Wegener, on the other hand, did not do his own field-work—at least not initially. Instead, he began with an initial hypothesis and then made a systematic but selective survey of the geological literature of the day, choosing facts and figures that supported his hypothesis and ignoring ones that did not.[35] This approach to formulating a theory was seen (particularly by North American geologists) as unscientific and one which they thought, in all probability, would yield a false result. Again, such a situation creates a problem for both Wegener's theory and the established (preferred) methodology. For instance, if Wegener's theory had been outstandingly successful at the empirical level, then this would tell against the accepted methodology. The methodological criticism was accepted by North American geologists as another principal reason to reject Wegener's drift theory.

In the early 1970s a similar methodological criticism was made against some theories which had caught the popular imagination. The idea behind these theories was that extraterrestrial beings visited ancient human societies and had a profound effect upon their later genetic, cultural and spiritual development. This methodological criticism was one of many which assisted in the ridicule and demise of these theories.

World-view

The difficulty here is a conflict between a scientific theory and some other strongly held, non-scientific belief. Such beliefs form part of an accepted view of the world and may be metaphysical, religious, ethical or something else entirely, provided they are strongly held.[36] A suitable example is the heliocentric theory of Copernicus. This theory is a scientific one. Despite being scientific, the theory was incompatible with the prevailing view of the world in the seventeenth century. The primary premise—that the Earth moved about a stationary and central Sun—was contrary to the religious belief that God had placed Man (and hence the Earth) at the centre of the Universe. In this view, the Earth must be the centre of all motion and itself motionless. This obvious conflict constituted an external conceptual problem for both Copernican theory and for Catholic theology (since the relation of inconsistency is symmetric). The Roman Catholic Church at that time was not prepared to revise its interpretation of the Scriptures and instead decreed that the doctrine of the Earth's motion was heretical.[37] This does show, however, that Laudan has a place in his theory of science for so-called 'non-scientific' influences.

Before concluding this section on conceptual problems, let's summarize Laudan's ideas on the weighting of such problems. He claims that as a general principle it is correct to say that a conceptual problem will carry greater weight than an empirical anomaly (although he does allow for the occasional exception). In terms of the cognitive relationships detailed earlier, he claims that the greater the magnitude of cognitive threat to a theory, the greater the weighting of the conceptual problems concerned. In a conflict between two theories (be they both scientific or only one), the weights of conceptual problems for one theory will depend on how acceptable the other theory appears. When we have two competing scientific theories in the same discipline and one theory generates conceptual problems not found in the other, then those problems take on a greater significance than if they were to be found in both theories. Lastly, Laudan claims that the weight of a conceptual problem increases with its age and the number of attempts made to resolve it.[38]

THE STRUCTURE OF EVOLVING RESEARCH TRADITIONS

Laudan's theory of science is a structured account bearing many similarities to the theories of both Kuhn and Lakatos. Laudan's equivalent of a Kuhnian paradigm or a Lakatosian scientific research programme is a 'Research Tradition'. A research tradition consists, among other things, of a series of individual theories that are responsible for solving different problems (both empirical and conceptual) at different times in a manner analogous to the succession and role of the series of theories found within Lakatos's research programmes. The three main features common to all research traditions are as follows:

> 1. Every research tradition has a number of specific theories which exemplify and partially constitute it; some of these theories will be contemporaneous, others will be temporal successors of earlier ones;
> 2. Every research tradition exhibits certain *metaphysical* and *methodological* commitments which, as an ensemble, individuate the research tradition and distinguish it from others;
> 3. Each research tradition (unlike a specific theory) goes through a number of different, detailed (and often mutually contradictory) formulations and generally has a long history extending through a significant period of time. (By contrast, theories are frequently short-lived.)[39]

The Laudanian research tradition performs much the same function as a Kuhnian paradigm or a scientific research programme in that it is a sort of 'containment vessel'. This vessel not only places limits on the nature and extent of the individual theories that it contains but also on what experimental methods are acceptable. A research tradition provides hints for researchers on how to construct theories in much the same way as Lakatos's positive heuristic does. The research tradition also defines the types of entities which are suitable for use in its component theories. In doing so, a research tradition excludes any entities which are not consistent with the decrees of the research tradition. In Laudan's words:

> *a research tradition is a set of general assumptions about the entities and processes in domain of study* [its allowed ontology], *and about the appropriate methods to be used for investigating the problems and constructing the theories in that domain* [its allowed methodology].[40]

Laudan presents an example of such ontology and methodology set in the geological research tradition of 'Uniformitarianism'.[41] The origin of this tradition is credited to Charles Lyell who set out its parameters in his *Principles of Geology*, published in 1830. Uniformitarian geologists believed that processes of the Earth were roughly cyclic, alternatively building up and wearing down surface features. On average, the Earth's geology could be said to be roughly uniform over long time periods. The ontology of uniformitarianism is essentially the presently observable causes of geological phenomena: for example, erosion, weathering, small-scale volcanic action, small-scale uplift and folding. The existence of these causes gave rise to all geological processes such as mountain building and canyon formation. The methodology of uniformitarianism was to apply these same causes to past situations in order to account both for the Earth as it is today and to explain traces left from past geological eras—fossils, strata, and so on. The catch-cry of the uniformitarian geologist sums up this methodology: 'the present holds the key to the past'.

It is important to realize that each research tradition has a beginning (and perhaps an end) but from its inception it develops, or more appropriately, it evolves. It follows that research traditions are best viewed in their historical context. The evolution of a tradition may result in development into a form which is totally different from its earlier ones. Laudan even includes the possibility of a research tradition developing over time into a version which is inconsistent with its earlier forms![42] How does this evolution take place? The most obvious way in which a tradition can change is by modification of its component theories. The role of these theories is to provide solutions to the empirical and conceptual problems which confront the tradition. Theory modification or even replacement within a tradition is one means by which such problems are solved. The replacement and modification of the constituent theories of a research tradition must, however, be done in accord with the ontology and methodology that define it.[43]

In the astronomical tradition referred to as heliostatic, for example, the constituent theories complied with the basic premise that all the planets (including the Earth) moved around a stationary Sun. The theoretical details were otherwise very different; for example, Copernicus had circular orbits complemented by minor epicycles, Kepler had elliptical orbits and Newton introduced perturbations ('wobbles') in the planetary ellipses. Copernicus's original formulation is, in a very real sense, inconsistent with the theories of Kepler and Newton. Yet this does not constitute a discontinuity in the tradition since they are linked both historically and conceptually. There is an historical evolution from the publication of Copernicus's famous treatise on planetary motions, *On the Revolution of the Heavenly Spheres*, in 1543 to Newton's *Principia* in 1687.

A research tradition does not entail its component theories; that is, its constituent theories cannot logically be derived from the premises of the tradition. A research tradition does, however, both limit the extent of the theories it contains and the range of problems its theories may address.[44] In accordance with its methodology, a research tradition offers researchers 'vital clues for theory construction' and acts as a guide (in a very similar manner to Lakatos's positive heuristic) for modification of theory with the aim of improving the ability of a theory to solve problems.[45] Quantum mechanics, for example, as it was originally developed by Schrödinger and Heisenberg in the 1920s, began to show unacceptable empirical discrepancies when used statistically to predict the motion and energy of an electron travelling close to the nucleus of an atom. Paul Dirac modified quantum theory to incorporate corrections required by the special theory of relativity whenever speeds close to the velocity of light are involved, as is the case with electrons that move close to an atom's nucleus. Dirac's resulting equation for the electron was highly successful in solving the empirical problem. This alteration of theory by Dirac was well within the limits imposed by the quantum physics research tradition and in accord with its (somewhat implicit) guidelines for theory modification.

A gradual change in the theories contained in any research tradition is only one form of evolution possible in Laudan's scheme. The assumptions, ontology and methodology of a tradition may also change. Unlike Lakatos's hard core, there is no unique set of premises that is held to be fundamental at all times in the life of a research tradition.[46] Changes in a research tradition as opposed to changes in its constituent theories will, in general, only occur when researchers find that modifications to these constituent theories do not remove the problems (be they empirical or conceptual) that are causing 'tension' for the tradition. In cases where changes to a constituent theory fail to resolve the research tradition's difficulties, the researchers concerned may try making small changes to the research tradition's ontology, its methodology or both.[47] If small changes to the tradition do lead to the resolution of some particularly troublesome problems, then this action will be judged as justified. Such action can solve the problems confronting a research tradition whilst retaining most of its assumptions. A research tradition can slowly evolve in a manner such that its appearance over a long time interval may be greatly altered, but at no stage in the life of the tradition are there any large-scale changes inflicted. Laudan writes:

> There is much continuity in an evolving research tradition. *From one stage to the next*, there is a preservation of most of the crucial assumptions of the research traditions. Most of the problem-solving techniques and archetypes will be preserved through the evolution. The relative importance of the empirical problems which the research tradition addresses will remain approximately the same. But the emphasis here must be on *relative* continuity between *successive* stages in the evolutionary process. If a research tradition has undergone numerous evolutions in the course of time, there will probably be many discrepancies between the methodology and ontology of its *earliest* and its *latest* formulation.[48]

Laudan denies that research traditions have hard cores in the Lakatosian sense—that there is a single set of basic assumptions in a particular research tradition that cannot be altered or rejected during the tradition's lifetime. The fundamental assumptions of a research tradition could perhaps be tagged a 'soft' core—one which is capable of being slowly moulded or changed to suit new circumstances, but a core nonetheless. (Laudan does use the words 'core' and 'central elements'.) This soft core is a set of elements or assumptions that are, at particular stages in the lifetime of a research tradition, taken as unrejectable by adherents to that tradition.[49] The elements comprising the soft core may be different at different times. A research tradition's core has continuity of existence because only gradual change occurs amongst its elements. This soft core is composed of the most central elements of a tradition and thus cannot be rejected without rejection of the research tradition itself. There is a subtlety here between the notions of changing the core and rejecting it, as Laudan explains:

> *at any given time* certain elements of a research tradition are more central to, more entrenched within, the research tradition than other elements. It is these more central elements which are taken, at that time, to be most characteristic of the research tradition. To abandon them is indeed to move outside the research tradition . . . I want to insist that *the set of elements falling in this (unrejectable) class changes through time.*[50]

An example of a changing core of a research tradition is found in the succession of continental drift theories. The core of Wegener's theory originally included the assumption that continents had ploughed through the more dense ocean floors but this assumption was absent in later versions of drift theory. Nevertheless there is not only an historical but also a cognitive continuity from Wegener's version to these later ones.[51]

Laudan asserts that all core elements of a research tradition may change over time. This could only be the case where the research tradition's defining criteria were not in explicit form—where there were no explicit core assumptions; instead the tradition was only known via its exemplars.[52] The assertion that all core elements may change is also not supported by historical evidence—there always appears to be at least one premise common to all successive theories in a single research tradition. For example, in all versions of heliostatic astronomy it is the Earth which is postulated to go around the Sun; in all versions of continental drift, it is postulated that the continents move laterally over the Earth's surface in geological-time intervals. In addition, Laudan does not specify how scientists should make decisions about which elements of a research tradition are unrejectable. How do the researchers decide which elements form a tradition's core?

PROGRESS AND THE ASSESSMENT OF RIVAL RESEARCH TRADITIONS

Laudan's fundamental unit of progress is a solved empirical or conceptual problem. He defines the aim of science as maximizing the number and scope of solved empirical problems, whilst minimizing the number and scope of anomalies and conceptual problems for one's own theory.[53] (Laudan calls this his 'mini-max' strategy.) Central also to Laudan's account of science is the appraisal of an individual theory in terms of its overall problem solving effectiveness:

> *the overall problem-solving effectiveness of a theory is determined by assessing the number and importance of empirical problems which the theory solved and deducting therefrom the number and importance of the anomalies and conceptual problems which the theory generates.*[54]

There are enormous difficulties inherent in this mini-max criterion. Firstly, in what pragmatic sense can it be asserted that one can count the number of empirical problems solved (for in principle this should be infinite) and the number of anomalies and conceptual problems generated? Secondly, if it did prove possible to count numbers of problems, then how are various weights (judgements of importance) to be assigned to particular problems? If we want to make a tally of the problems in one research tradition and make a comparison with a similar tally from another research tradition, then it is not sufficient to merely arrange the problems in each research tradition in their order of importance. We would need to assign some numerical value to each problem in order to indicate its weighting, but how is this to be done? Laudan does not provide a method.

Progress, for Laudan, occurs when the successive theories in a research tradition increase in problem-solving effectiveness over time.[55] Yet Laudan presents more than one measure of progress and two modes of evaluation. Consider firstly the adequacy of a research tradition: this is a rating based on the level of problem-solving effectiveness of the theory or theories current in a research tradition, without reference to any previous theories. This can only be a measure of how effective a tradition is at solving problems at present.[56] (Any research tradition may, of course, get better or worse at solving problems.) The level of adequacy of a research tradition is arrived at by evaluating the problem-solving effectiveness of each current theory in the tradition and adding these individual effectiveness ratings.

The second mode of evaluating a research tradition is called progressiveness. This mode necessarily involves comparison between the more recent theory or theories in the tradition and earlier ones. This must be the case if an assessment is to be in accord with the defined notion of progress as an increase in problem-solving effectiveness over time. The general progress of a research tradition is found by comparing the adequacy of the original theories in the tradition with the adequacy of the current theories.[57] What Laudan calls the rate of progress is different again. The rate of progress of a research tradition indicates changes in the adequacy of the tradition during a given time interval. Now these measures need not and usually will not be the same. Something else is needed to complete the evaluative circle.[58]

The next part of Laudan's theory concerns the contexts in which theories and research traditions are evaluated. It is an important part of Laudan's theory that research traditions are not evaluated in an absolute sense; that is, their evaluation is done with respect to other existing research traditions in the same discipline. He holds that it is the usual case in science for there to be a plurality of competing theories or research traditions rather than a single dominant theory or paradigm. As previously noted, the historical evidence supports the view of plurality. Laudan suggests that there are two different contexts for the evaluation of research traditions and their theories: 'Acceptance' and 'Pursuit'. What does he mean by 'acceptance'? If scientists accept a theory or research tradition, then they treat it 'as if it were true'.[59] How is such a choice to be made? What criterion can there be for accepting one research tradition instead of another? Laudan's answer is to choose the research tradition with the greatest problem-solving adequacy. Furthermore, such a choice is, for Laudan, fully rational.[60]

The second context of evaluation is 'pursuit'. The picture of appraisal is incomplete, argues Laudan, without a notion of exploration for theories that are not accepted but are nonetheless worked upon. Laudan's reasons for including a second context are twofold: historical cases show that scientists often actively pursue (that is, explore or use) a research tradition different from the one which they accept, and the activity of scientists would be largely inexplicable in rational terms if the context of pursuit were ignored. The question obviously arises—which research tradition is it rational to pursue? Laudan's answer is one which has a greater rate of progress than its competitors.[61] As with the criterion for acceptance, any evaluation of which research tradition to pursue is made on a comparative basis.

The specified criterion for pursuit of a research tradition—its having the highest rate of progress—is not specific enough. Recall that the rate of progress is indicated by changes in the adequacy of the tradition during a specific time interval. But what time interval should this be measured over?[62] It is not the time interval since the inception of the tradition, which Laudan explicitly calls 'the general progress'. Is the time interval required to measure 'rate of progress' to be long or short? It cannot be too long for then it would tend to the 'value' of the general progress. It cannot be too short, for then it would tend to the 'value' of the (current) adequacy. Laudan needs to provide a way of determining the length of this time interval and in what stage of the tradition it falls. Laudan also does not say how acceptance of a research tradition may be distinguished in practice from vigorous pursuit, as the same activity is required for both.[63] Indeed, he gives no compelling reasons why a researcher has to accept one theory and pursue another, rather than just pursuing several.

Before proceeding any further, let's clarify and expand on some of the points already presented. In the introductory section, Laudan was quoted as saying that he was troubled by philosophers of science making 'progress parasitic upon rationality'. What Laudan has done is effectively to turn this around. Instead of making it progressive to choose the most rational research tradition, he presents the case that it is rational to choose the most progressive research tradition, where 'most progressive' is defined as having the highest problem-solving effectiveness. But

Laudan offers even more than this—he presents an evaluative context for rival research traditions not previously afforded by theories of science and scientific change. In comparison, what do Kuhn and Lakatos offer in regard to paradigm or research programme choice? Kuhn and Lakatos specify only two cognitive stances: *acceptance* or *rejection.* Laudan originally offered three stances: *acceptance*, *pursuit* or *rejection.* Indeed, Laudan later widened the evaluative context. Instead of merely having acceptance, pursuit or rejection, he advocates a spectrum of possible responses from complete acceptance, through different degrees of pursuit, to complete rejection:

> The logic of acceptance and rejection is simply too restrictive to represent this [historical] range of cognitive attitudes . . . My view is that this continuum of attitudes between acceptance and rejection can be seen to be functions of the relative problem-solving progress (and rate of progress) of our theories.[64]

Laudan does not reject the notion of the incommensurability of scientific theories but he does argue that Kuhn's case for the *radical* incommensurability of theories is incorrect in several respects. Although not denying that there could be cases where theoretical terms from different theories could not be intertranslated or translated into some theory-neutral language, Laudan claims that objective and rational comparisons can still be made.[65] He offers two arguments to support this view.

Laudan's first method of objective comparison is appropriate where competing research traditions solve the same problems. This, of course, requires that it is indeed possible to show that rival traditions tackle at least some of the same problems. He claims that this is possible provided that the low-level assumptions which are used to specify a problem are not inconsistent with each of the competing research traditions.[66] (Recall that low-level assumptions are those containing minimal theoretical content and which are implicit in the most basic 'observational' statements. See pp. 17–18.)

An example may clarify this argument. Since the time of the ancient Greeks, mathematical astronomers have attempted to solve the 'problem of the planets'. The problem is to explain and to predict accurately the paths of the planets when they seem (temporarily) to reverse their direction of motion through the 'fixed' background of stars. The low-level assumptions required by each of the theories which attempted to explain this phenomenon of planetary retrogression were not inconsistent with those theories. Within each astronomical theory it is possible to state the problem in a manner that does not contradict competing theories and that has meaning for the adherents of each of the different theories. For instance, the direction of motion of a planet with respect to a given set of fixed stars at a particular date can be unambiguously specified. So regardless of the mechanism that a single astronomical theory employed in its explanation of motion (for example, homocentric spheres, geocentric circles, heliocentric circular orbits, heliostatic elliptical orbits, and so on) all astronomers agreed on what the problem was. Thus Laudan can claim that problems shared in this way provide a basis for rational appraisal.[67].

Laudan's second approach to objective assessment of rival and perhaps formally incommensurable research traditions is to compare (by a tabulation or similar method) the progressiveness of each individual research tradition. Recall that the progressiveness of a research tradition is ascertainable within the tradition itself. Laudan claims that a progressive ranking of research traditions could be achieved even if the traditions are incommensurable.[68]

RATIONALITY, SCIENTIFIC REVOLUTIONS AND DEMARCATION OF SCIENCE IN LAUDAN'S THEORY

In Laudan's theory, rationality is presented as a relative concept. The norms (or canons) of rational decision-making are not absolute but are relative to time and place. Thus Laudan would reject the basic assessment criterion for the rationality of an action or choice presented earlier (see p. 52) as too simple and in need of supplementation. He writes:

> any appraisal of the rationality of accepting a particular theory or research tradition is trebly relative: it is relative to its competitors, it is relative to prevailing doctrines of theory assessment, and it is relative to the previous theories within the research tradition.[69]

Yet one feels inclined to ask what is wrong with a notion of rationality that declares a rational action as one that offers the greatest promise of achieving one's aim? Clearly, there is nothing, in itself, wrong with such a notion. If the aim of science is to solve problems, why isn't this notion of rationality sufficient if it requires the researcher, as it should, to merely accept the research tradition with the greatest problem-solving adequacy? Is this not sufficient to achieve the stated aim and without recourse to other factors? This would be the case, except that stating the aim of science according to Laudan as just solving problems is to misrepresent him. The aim of science according to Laudan is expressed correctly by his mini-max strategy, namely, to maximize the number and scope of empirical problems solved by a research tradition while minimizing the number and scope of anomalies and conceptual problems for that tradition. It is within the framework of this strategy as the aim of science that an assessment of whether a choice is rational or not becomes relativized. Let's examine each of the three relativities of assessment mentioned by Laudan to see why.

Laudan claims the choice of which research tradition to accept from a selection of rivals is to be made on the basis of the highest problem-solving adequacy. This means having to make a comparison of the respective adequacies of rival traditions. If rationality consists in choosing to accept the most adequate research tradition from a set of competitors, each of which is currently identified by a theory (or set of theories) that is the latest in a historically evolved series, then a judgement of the rationality of such a choice is inescapably bound up with the factors that govern the problem-solving assessment of individual theories. It is, of course, one thing for the rational choice to be intimately linked to theories within a research tradition and to other research traditions in the same discipline, but quite another for the choice to be linked to factors external to the relevant scientific discipline. Laudan makes this connection through the existence of external conceptual problems. External conceptual problems weigh very heavily in the assessment of a research tradition's acceptability. In this way the influences of extra-scientific effects can be shown to have a rational basis.

Let us consider the following example. After the popularizing of Copernicus's heliocentric theory by Galileo and its subsequent denunciation by the Roman Catholic Church in 1616, Catholic astronomers could not officially hold to or use the Copernican theory.[71] In 1588 Tycho Brahe published the details of a geocentric (Earth-centred) system in which the Sun revolved about the Earth but all the other planets revolved about the Sun. The Tychonic system was geometrically equivalent to the system of Copernicus.[72] It therefore had all the empirical advantages of the Copernican system with the added bonus that its acceptance did not cause any 'conflict'—there was no external conceptual problem as Tycho's system was consistent with a central and motionless Earth. In terms of the prevailing opinion of the time, it was indeed rational to accept the Tychonic system over the Copernican. The former was more progressive because it did not suffer from the external conceptual problem of the latter.

The notion of rationality given by Laudan does not seem to account for a situation where, although the majority of scientists accept the research tradition with the greatest problem-solving adequacy, some individual researchers accept a research tradition that does not have the greatest problem-solving adequacy. This would not be a problem for Laudan's model if it never happened, but there are instances in the history of science where this has occurred. The problem-solving model's incapacity to account successfully for rational, individual choices that go against the majority decision is one rather acute failure.

We turn now to consider Laudan's definition of a scientific revolution. Many historians of science and some philosophers have over-exaggerated their descriptions of those periods in the historical development of science known as 'scientific revolutions'. Laudan's opinion on the matter is that these 'revolutions' are much milder than Kuhn (and others) have made them out to be. In Laudan's model of science there is constant rivalry and debate, rather than long periods of normal science interspersed with the occasional revolution.[73] Laudan argues that (*contra* Kuhn) a revolution is not achieved by the conversion of most scientists to a new paradigm (or research tradition). Instead, when revolutions do occur they are effected by a small number of scientists. Moreover, a revolution occurs not so much when the members of a scientific community abandon their previous research tradition in favour of a new one, but rather when that new tradition makes such (rapid) progress that it cannot be ignored. When this situation arises, the researchers in the field may respond by accepting the new tradition, by pursuing it, or at the very least, by modifying the older tradition.[74]

The Chemical Revolution in the eighteenth century, for example, was secured by a handful of French scientists led by Antoine Lavoisier. The tactic employed was first to introduce a new, systematic nomenclature for chemistry which implicitly embodied many theoretical assumptions of Lavoisier's oxygen theory. The clarity and consistency of the new nomenclature found it wide acceptance. Lavoisier followed up the publication of his nomenclature with what is today considered a classic textbook on his chemical theory, *Elements of Chemistry*, in 1789. Lavoisier's theory meshed very neatly with the newly accepted nomenclature. Many of the older chemists of the time, who stood by the existing phlogiston theory of combustion and calxification, were forced to take notice of and respond to Lavoisier's work as it had become a major cognitive threat. Laudan would claim that it was at this stage—that is, when other researchers in the field felt compelled to pay serious attention to Lavoisier's theory—that the Chemical Revolution occurred.

The last topic in this section concerns Laudan's thoughts on the nature of scientific knowledge and the science–non-science demarcation. The problem-solving model does not require theories to be true or even approximately true. Scientific theories simply have to solve scientific problems. Scientific knowledge, in Laudan's terms, is not a steady accumulation of fact and theory but is by its very nature non-cumulative. A change of research tradition carries with it new ontologies and a new methodology to the exclusion of the old. This gives Laudan his own version of 'Kuhn Loss'. Laudan's view on demarcation is summarized as follows:

> The approach taken here [the problem-solving model] suggests that there is no fundamental difference in kind between scientific and

> other forms of intellectual inquiry ... The quest for a specifically scientific form of knowledge, or for a demarcation criterion between science and non-science, has been an unqualified failure.[75]

Thus Laudan sees no basic difference between science and other academic pursuits. He does acknowledge that science is more progressive than other fields, but he accounts for this only as a difference of degree and not of kind.

The question now to be posed is whether Laudan's theory has the essential ingredients of a theory of science and scientific change. Clearly some commentators, such as Paul Feyerabend, think very little of Laudan's efforts.[76] Feyerabend is in the minority though, at least as far as philosophers are concerned. Others commentators, such as Le Grand,[77] find much importance in Laudan's theory and its ability to make sense of many more historical instances of theory choice and theory change in science than could the theories of Kuhn and Lakatos. The criticisms presented in this chapter indicate that there still remains much to be done in order to construct a theory which encompasses more aspects of science and accounts for more of the detailed history of science.

5
The Sociology of Science

There are many accounts of scientific research in terms of the social environment in which that research is conducted. It has become conventional to split the social study of science into two categories: the 'Sociology of Science' and the 'Sociology of Scientific Knowledge'. The sociology of science (more properly called 'Non-cognitive Sociology of Science') is primarily concerned with the social network and hierarchy of scientists in institutionalized science. The (non-cognitive) sociology of science is not concerned with social influences on scientific methodologies or the status of knowledge claims by scientists. These latter considerations are in the realm of sociology of scientific knowledge (also called the 'Cognitive Sociology of Science'). In this chapter we shall deal exclusively with non-cognitive sociology of science.

THE SOCIAL STRUCTURE OF SCIENCE

One of the pioneers in the study of the sociology of science was the American sociologist Robert K. Merton, whose relevant papers from the 1940s and 1950s have been collected in one volume.[1] Merton is seen by many as the 'founding father' of sociology of science. The writings of Merton have been supplemented by other sociologists and they have tended (until recent times) to focus their attention almost exclusively on the institutional structure of scientific research; that is, on the social organization of the scientific community and the rules by which this community conducts its activities. Such analysis is termed 'functionalist' as it explores the functions of these social rules. The attention of Merton and similar sociologists was not directed at the knowledge claims of science. Merton accepted the view that scientific knowledge was generated by objective methods and as such reflected regularities of nature.[2] Scientific knowledge itself was therefore thought as being beyond the scope of sociological analysis. The job of explicating objective scientific methods was left to philosophers of science, particularly the logical empiricists.

Merton also developed strong views about how different social contexts affected the development of science. He came to the conclusion that although many diverse cultures and societies have provided support for science, it was western democratic countries that offered the greatest opportunity for development of science and totalitarian societies that most restricted its advance.[3]

If the workings of science are to be understood as arising from some kind of social organization, then let's begin with a straightforward definition of what constitutes a social organization. At a basic level of analysis, it is plainly obvious that scientific research is conducted and managed by people (not machines!). People exhibit different behaviour patterns in different circumstances. The relevant behaviour patterns are those of people actively engaged in scientific research. All social bodies possess certain common traits and this includes scientific ones. In regard to these traits we shall use Norman Storer's definition of a social system:

> a social system [is defined] as a stable set of patterns of interaction, organised about the exchange of a qualitatively unique commodity and guided by a shared set of norms that facilitate the continuing circulation of that commodity.[4]

Under this definition the commodity in relation to science is certified knowledge, that is, 'empirically confirmed and logically consistent statements of [natural] regularities'.[5] The aim (or goal) of institutionalized science is the extension of certified knowledge.

In order to achieve this goal of certified knowledge and to regulate its own activities, the social system of science is governed by a number of institutional imperatives, also called mores. (In his definition, Storer refers to these imperatives as a shared set of norms.) Merton defined four institutional imperatives for the conduct of science. He arrived at these after making some sociological studies of the history of science. His study of the interaction of science and society in seventeenth-century England proved especially useful. Merton writes:

> the ethos of science . . . can be inferred from the moral consensus of scientists as expressed in use and wont, in countless writings on the scientific spirit and in moral indignation directed towards contraventions of the ethos.[6]

Merton's four institutional imperatives are as follows:

(1) 'Universalism'. This is a principle whereby the theories and laws of science are accepted on a basis that is independent of the social and personal characteristics of their discoverers or originators. Scientific knowledge is objective and therefore must be impersonal.

(2) 'Communality' (or 'Communism'). This imperative states that the scientific commodity is not personal property but belongs to the scientific community as a whole. Recognition of priority of discovery is granted to individuals but the knowledge as such becomes common property of all scientists. Consequently individuals or groups of scientists who make discoveries are obliged to communicate their results to the scientific community as a whole.

(3) 'Organized Scepticism'. Each scientist is responsible to ensure that the basis of his or her research (especially when that basis is due to others) is correct. More generally, all beliefs relevant to one's research must be scrutinized in terms of logical and empirical criteria. All published research must also be subject to a high level of criticism by the scientific community to ensure objectivity. (Compare this with Scheffler's account of the standard view of science, pp. 9–11.)

(4) 'Disinterestness'. The scientist should be sufficiently detached from his or her research to avoid circumstances of personal profit arising from that research; for example, the scientist should avoid the pursuit of recognition as a goal in itself or for monetary benefit. Although individual scientists may have many different motives for pursing particular pieces of research, it is in the general interest of the scientist to

conform to the institutional norms of science and to pursue the goal of certified knowledge.[7]

In addition to these four imperatives, there have been a number of others added by various writers on the sociology of science at different times. Some of these additional norms could have been subsumed under Merton's original four and some add new content. We will only concern ourselves with two further norms of science, proposed by Bernard Barber:

(5) 'Rationality'. This norm affirms a belief in the rationality of scientific methods and that scientific research yields certified knowledge. Such belief sustains the efforts of scientists in the face of the extreme problems that occur from time to time in their research.

(6) 'Emotional Neutrality'. The emotional neutrality of a scientist requires that a researcher not be so involved in his or her research (or favourite theory) that she or he will not try a new approach or reject a theory if empirical evidence so dictates. Emotional neutrality is seen as a condition for achieving rationality.[8]

The above six norms are postulated as the essential rules and moral principles by which the scientific community regulates itself. Important to the successful functioning of these norms are processes of critical evaluation and peer review.

THE REWARD STRUCTURE OF SCIENCE AND ADHERENCE TO ITS INSTITUTIONAL NORMS

Merton claims that the institution of science has a system for the allocation of rewards to scientists for conforming to its institutional mores.[9] These norms exert a certain social pressure on researchers to make original contributions to the body of scientific knowledge; the researcher must not only make scientific advances but must also communicate these findings to the scientific community. There are many rewards for service to science. Two of the most important rewards will be considered: recognition and promotion.

Recognition comes in various levels. It may be as simple as having one's name included in a published scientific paper.

Alternatively, recognition may take the form of *eponymy*—where a process, theory, equation, physical constant, object, or branch of science is named after its discoverer: for example, Boyle's Law, Darwin's Theory, Einsteinian Relativity and so on. Merton described the reward of recognition as 'socially validated testimony [in] one's role as a scientist'.[10] Under the imperative of communality, the scientific commodity cannot be individually owned. The only allowed form of 'property rights' is having one's name affixed to some particular scientific commodity, for example, a virus, and this 'right' is only conferred upon the first researcher to make public his or her discovery. Eponymy is seen as one of the highest rewards in science. The convention of bestowing the name of the researcher who first makes a particular scientific commodity public knowledge has led to both competition amongst researchers working on the same problem and to many disputes of priority of discovery.

The classic example of a priority dispute was the clash between Sir Isaac Newton and Gottfried Wilhelm Leibniz. Newton was first to invent what he called 'Fluxions' (now referred to as the 'Differential Calculus') in 1666 but he kept its discovery a secret. Leibniz independently developed his own version of the calculus in about 1676 and published it in 1684. What followed was a long-running, unpleasant confrontation over who had first made the discovery.[11] The history of science is littered with similar priority disputes.

The other type of reward which we shall briefly discuss is promotion. In institutionalized science all researchers (be they employed by government, tertiary institutions or private enterprise) hold positions of a particular rank or level. In the course of their working lives, all researchers could expect to gain several promotions to higher levels on the basis of their research performances. Promotion, therefore, can be seen as a far commoner form of reward for successful research than eponymy. The rate of promotion and final level reached will be a function of how successful the individual's research endeavour is judged to be. Promotion will, in general, depend on a researcher's work gaining at least some form of recognition. The normal manner in which this occurs is by peer review. For example, the decision to publish a scientific paper is taken by journal editors on the advice of referees who are themselves research scientists. In this way a judgement on the value of an individual scientist's research is made by his or her peers. A favourable judgement is a form of recognition for work which is judged as valuable to science. Continued recognition and publications ultimately lead to promotion. Promotion brings with it a higher social status and greater prestige for the researcher, in addition to an increased salary.

Two questions arise out of the preceding discussion. What makes the scientist conform to the institutional norms of science? What part does the reward structure play in gaining the adherence of scientists to the norms? Merton does not really provide answers to these two questions.[12] Another sociologist, Warren Hagstrom, offers a fairly blatant response—he claims that the outstanding motivation of scientists is simply to gain recognition. In this view, a scientist receives recognition in exchange for contributing certified knowledge to the scientific community. Adherence by scientists to the institutional norms is assumed to provide the optimal way of achieving certified knowledge which, in turn, can be exchanged for recognition. Hagstrom therefore concludes that adherence to the norms is (by a majority of scientists) assured by their desire for recognition.[13] Although this response does provide answers to both the above questions, Hagstrom's analysis ignores other legitimate motives. These include an honest curiosity about natural phenomena, an altruistic desire to contribute something lasting to humanity, or just the sheer enjoyment of solving scientific puzzles.

Norman Storer, by contrast, has included as a motive the urge in scientists to be creative and to complete successfully all or part of a given research endeavour. He argues that adherence to the norms of institutional science is seen by researchers as the best way of continuing the fair allocation of the scientific commodity. Recall that under Storer's definition of a social system, interactions in the system are dependent on the exchange of a unique commodity, in this case, scientific knowledge. In order for everyone in the social system of science to receive their just share of the scientific commodity, there is implicit agreement amongst the members of the community to abide by certain rules, namely, the institutionalized norms. In other words, if everybody plays by the rules then everybody gets a fair go! This is seen by the members of the community as the best way to optimize their chances of being successful in science, reaching its aim and being rewarded by doing so. Thus the source of the moral force (as Storer calls it) to adhere to institutional imperatives can, in his opinion, be shown to be derivative from a general analysis of a social system where the basis of the system is the scientific commodity.[14]

Whereas Merton had said that the rewards of science and the imperatives are separate consequences of scientists' devotion to advancing knowledge, Storer claims the imperatives are derivative from the scientist's interest in the rewards of science.[15] Storer's account is not identical to that of Hagstrom, for he includes in his description of 'reward' an additional item to recognition—completion of the creative act. This form of reward is taken by Storer to be as important to the scientist as the gaining of recognition. Storer's argument is summarized in the following passage:

> scientists subscribe to the norms of science first of all because of their importance for the continued, adequate circulation of the commodity in which they are mutually interested . . . It is the occasional reinforcement given these norms by the scientist's awareness to their relevance to his own interest in obtaining competent response to his work rather to the general goal of science, which I feel accounts for their continued moral potency . . . these norms are natural concomitants of the desire to be creative.[16]

Inclusion of the creative act and the desire to complete it does add credence to this analysis of the behaviour of the scientific community. At this point it is perhaps appropriate to consider what the available evidence indicates about these social accounts of science. Case-studies of several different research groups show that there are statistically significant departures (in both number and kind) from the six stated institutional norms.[17] The results of these studies therefore cast serious doubt upon the validity of the above institutionalized pictures of science.

DEVIATIONS FROM INSTITUTIONAL IMPERATIVES

The institutional imperatives of science should be seen, at best, as describing an ideal situation. When and if they do operate, it is only in periods of relative peace since wartime governments impose contrary rules on the scientific community: for example, secrecy, disinformation. When the imperatives can be seen to be (mostly) operating, science is conducted in a manner resembling Kuhnian normal science. Even in the course of routine research the following deviations from institutional imperatives have been observed:

- theories which originate in particular countries, for example, Chile, South Africa, have been immediately suspect as to the legitimacy of their content. In other cases, the personality, social standing, criminal record and so on of individuals who have proposed theories have influenced the reception and content appraisal of those theories. Both of these judgemental stances are contrary to the norm of universalism as it holds that the origin of a scientific contribution should have no bearing on its assessment.
- scientists of long standing and demonstrated ability in a given field are not subject to as rigorous checks on their research as are new scientists. This goes against the principle of organized scepticism which holds that all research should be subject to the same high level of scrutiny and against universalism since the personal standing of the scientist should be irrelevant.
- concern over the gaining of suitable recognition has lead to situations where results are withheld until such time as their release will result in a greater impact. Theories have also been withheld from the public domain if they have immediate practical applications until such time as a suitable patent application has been lodged. The withholding of theories or experimental results is contrary to the norm of communality. The gaining of patents for the purposes of profit goes against the norm of disinterestness.
- research goals are, on a regular basis, specifically profit-oriented, in contradiction of the disinterestness imperative.
- in many cases a scientist has held on to a favourite theory for periods in excess of what could be considered reasonable in the

light of the available evidence. This goes against the principle of emotional neutrality.

- new, radical theories regularly receive adverse reactions because such theories strongly challenge widely accepted views. This is, again, contrary to the norm of emotional neutrality.

How are deviations from the institutional norms to be accounted for if they are statistically significant? Ian Mitroff attempted to answer this question after conducting a comprehensive study of Apollo Moon scientists. Mitroff inferred from his data (following on from an earlier suggestion of Merton's) that for each original institutional norm there exists an opposite one: a counter-norm. (Note that the original norms to which Mitroff refers are not just Merton's original four, but are an expanded set of eleven norms. We shall consider only those norms and counter-norms which are relevant to the present discussion.)[18] Mitroff proposed that there are (at least) two sets of institutional mores of science, namely, the norms and the counter-norms. The existence and operation of a set of counter-norms were postulated to explain what were previously characterized as deviations from institutional imperatives. Each of the counter-norms justifies action contrary to its corresponding norm.

The six relevant counter-norms of science proposed by Mitroff are:

(1) 'Particularism'. This counter-norm justifies such actions as taking the personal or professional attributes of a researcher into account when evaluating a scientific contribution from that person. This can be seen as opposite to the norm of universalism.

(2) 'Solitariness' (or 'Secrecy'). Opposing the norm of communality is one of secrecy. This holds that the property rights of the discoverer of a piece of scientific knowledge should include how and to whom this knowledge is transmitted. This control can only be maintained if the knowledge is not made public.

(3) 'Exercise of Judgement'. The norm of organized scepticism is somewhat balanced by the exercise of judgement. Scientific contributions should be gauged using logical and empirical

criteria only, according to the prescriptions of organized scepticism. The exercise of judgement legitimizes evaluation based on expert opinions of experienced scientists as to the worth of particular scientific contributions, without recourse, necessarily, to the above-mentioned criteria.

(4) 'Interestness'. This counter-norm asserts that scientists can work towards a personal or a specialized goal rather than the scientific community's goal of extending certified knowledge.

(5) 'Non-Rationality'. Mere faith in the rationality of scientific methods cannot guarantee that certified knowledge will always result from employment of such methods. Scientists (like all people) do not always act in a fully rational manner and scientific advances can result from non-rational as well as rational actions. Therefore, on some occasions, actions other than those which appear fully rational may be justified.

(6) 'Emotional Commitment'. A certain amount of commitment to a theory is essential to the advancement of science, contrary to the norm of emotional neutrality.[19]

The evidence presented by Mitroff is consistent with his espoused view but this is not the only interpretation possible. Michael Mulkay offers such an alternative interpretation. He begins by making the important point that social norms are only institutionalized if the relevant social institution provides rewards for continuing conformity to its norms and enforces penalties for non-conformity. He further claims that there is little evidence to suggest that rewards to scientists are actually made on the basis of their conformity to Mitroff's two sets of norms.[20] Therefore Mulkay reasons that these norms do not constitute fixed institutional rules and values. Instead he argues for a different interpretation on the basis of the evidence cited by Mitroff and other separate case-studies. Such evidence does indicate that scientists do describe their own conduct and that of other researchers in terms of Mitroff's two sets of norms, but their use in a descriptive role in no way establishes these sets of norms as being actual institutionalized rules.

Mulkay labels the two sets of norms as mere standardized

verbal formulations which scientists employ as a vocabulary 'to categorise professional actions differently in various social contexts'.[21] In other words, scientists only pay lip-service to the norms in order to describe their activities. So although there are no institutionalized normative principles to which scientists conform, none the less standard formulations of norms (such as Merton's or Mitroff's) form the basis of a larger, more flexible ethos by which individual scientists characterize their actions and those of their colleagues. Mulkay's argument is summarized in the following quotation:

> I am simply arguing that within the relatively distinct community concerned with scientific research, as indeed in most areas of social life, interaction cannot be adequately depicted as expressing any one or more sets of institutionalised normative principles or operative rules deriving from such principles. It seems more appropriate to portray the 'norms of science', not as defining clear social obligations to which scientists conform, but as flexible vocabularies employed by participants in their attempts to negotiate suitable meanings for their own and others' acts in various social contexts . . . what is clear *is* that it is highly misleading to regard the diffuse repertoire of standardised verbal formulations as the normative structure of science or to maintain that it contributes in any direct way to the advance of scientific knowledge.[22]

Mulkay's interpretation is difficult to ignore. If the institution of science itself fails to police the so-called norms of science, then these rules and moral imperatives are not in any real sense institutionalized or binding on members of the scientific community. Consequently any conclusions regarding the conduct of research scientists which are dependent on the premise of an institutionalized nature for scientific norms must be considered highly suspect. This then suggests that the social analyses presented earlier, be they based purely on institutional imperatives or on more general dynamics of social systems, simply fail to explain the actions of researchers. The wider picture of science depicted by Mitroff also suffers from a particularly damning criticism—which seems implicit in the above quotation from Mulkay—namely, if one is at liberty to postulate almost as many norms and counter-norms as one wishes then it is possible to explain *any* kind of behaviour!

A scientist's primary motivation, according to Mulkay, is self-interest (or interest in one's own research group). The verbal formulations chosen by scientists in order to justify the decisions they take will depend on particular social situations but where possible they will be such that the individual or group interests are protected. Mulkay also argues that these verbal formulations are constructed *post hoc*, that is, after the decisions are taken and in such a manner that the whole process appears to be dictated by rational norms rather than by self-interest.[23]

Consider again the example of the 'vis viva' controversy which was outlined on pp. 57–8. The relevant part of this controversy took place in the early 1700s between two rival groups of physicists over what constitutes the force of a moving body. The reactions of British Newtonian physicists to opposing views very clearly demonstrates that this scientific group also acted as a social grouping (or social community). The Newtonians acted to protect their own interests by affirming the correctness of their position and discrediting the views of their opponents. The most prominent Newtonian spokesman, Samuel Clarke, was even driven to such extreme measures as insulting those who held contrary opinions on the matter.[24]

Mulkay's characterization of science, although better than the others surveyed in this chapter, is still far from being a full explanatory account of the social influences on the scientific endeavour. We need to look somewhat deeper into the effects of socialization on scientists and on the theories they produce. This will be undertaken in the next chapter. We may also ask why it is that science is so successful if the motives of scientists are merely selfish. The question of the success of science will be addressed in Chapter 7.

6
The Sociology of Scientific Knowledge

RATIONALISM, RELATIVISM AND SOCIOLOGY OF KNOWLEDGE

What is sociology of knowledge? Most sociologists would not even attempt to offer a definition. Those definitions that do exist are extremely wide in their scope. Consider one such definition:

> The term "knowledge" must be interpreted very broadly indeed ... But whatever the conception of knowledge, the orientation of this discipline [the sociology of knowledge] remains largely the same: it is primarily concerned with the relations between knowledge and other existential factors in the society or culture.[1]

We have, up to this stage, not explicitly defined what knowledge is, tending instead to assume that its meaning would be either sufficiently well known or implicit in the content of earlier chapters. The broad interpretation of 'knowledge' required for the above definition is a far cry from the traditional philosophical one. Traditionally knowledge has been taken to be *justified, true belief.* However, this traditional view of knowledge cannot be sustained, for as Bertrand Russell (amongst others) has convincingly argued, all knowledge is tainted with doubt to some extent.[2]

When sociologists first turned their attention to science they assumed that scientific knowledge was certain knowledge—that it had a firm basis in the natural world. Scientific knowledge was justified by virtue of the 'The Scientific Method'. Facts and relations about the external world were not open to manipulation, they thought, and it is from these that scientific theories are inferred. Scientific knowledge was taken as being beyond reproach in the sense that social factors could play no role in the content of theories arrived at by the proper method.

In the 1950s, sociologists were beginning to change their minds about the claimed objectivity of scientific knowledge. It was not until the advent of T. S. Kuhn's *The Structure of Scientific Revolutions* in 1962, however, that wholesale sociological study of the form and content of scientific theories became legitimized. Why was this the case? Kuhn had cast doubt upon older characterizations of science, especially the standard view of science. He claimed, for example, that science was non-cumulative and that the highest standard was not something objective but merely the consensus of the relevant scientific community. Kuhn's theory, though welcomed by sociologists, was initially received with great consternation by many philosophers of science. Scheffler, for example, wrote the following depressing description:

> Paradigms, for Kuhn, are not only "constitutive of science"; there is a sense, he argues, "in which they are constitutive of nature as well."
>
> But now see how far we have come from the standard view. Independent and public controls are no more, communication has failed, the common universe of things is a delusion, reality itself is made by the scientist rather than discovered by him. In place of a community of rational men following objective procedures in the pursuit of truth, we have a set of isolated monads, within each of which belief forms without systematic constraints.
>
> I cannot, myself, believe that this bleak picture, representing an extravagant idealism, is true.[3]

Two primary considerations lead to the conclusion that scientific knowledge need not be true or objective: the theory-ladenness of observation and the underdetermination of theory by data. The realization that all observations were to some extent theory-laden meant that there could be no such thing as a set of purely objective scientific facts (compare the discussions on pp. 12–15 and pp. 16–21). The assumptions implicit in making any observation depend to some degree on the cultural and social context in which the observer lives. Even at the most basic empirical level the 'facts' of science cannot be relied upon as totally objective indicators of natural regularities, or as being unimpeachable evidence for the correctness or otherwise of scientific theories.

In Chapter 1 we saw that the underdetermination of theory by data necessitates the use of a methodology in making the choice between rival theories. What should determine the choice of one's methodology? Different methodologies are urged by different schools of philosophy and sociology of science and for their own particular reasons. The choice of methodology would, therefore, appear to depend on a number of different factors, including perhaps social ones. These considerations, regarding the basis upon which science is developed, open the door to sociological investigations of the causes and methods by which the content of scientific theories is arrived at and to what extent this content is socially (as distinct from empirically or rationally) determined.

What does it mean to say that the content of a theory is *socially determined*? Every human society, past or present, civilized or primitive, influences the manner in which its citizens perceive things and the ways they think about the world around them. This is not surprising, as everyone in any society learns both by tacit and by more obvious methods from those who have come before them. We all learn to speak, to act and to reason from our forebears. All societies exert pressures on their citizens to conform to established rules, beliefs and values. These pressures can be overtly imposed, for example, by harsh laws which are physically and ruthlessly enforced. Alternatively the pressure to conform may be entirely psychological, for example, where a particular sort of behaviour is adhered to because of concern about how one may be judged by others. The history of the world is a history full of such pressures imposed upon individuals and populations in order to make them agree to, or at least conform with, certain norms or beliefs. The more obvious examples are the many and varied political repressions and religious persecutions which have occurred over the ages.

The kind of influence which affects theory generation is usually much more subtle than explicit threats of physical violence or terms of imprisonment. The formation of hypotheses and their development into fully fledged theories depends both on the social context and the existing background knowledge. These place limits on the types of theories which can be developed in a particular society since concept formation will depend on what ideas already exist and on the sort of concepts that are consistent with existing patterns of abstraction. Consider some examples. In modern society, we have an elaborate explanation for the process of human procreation—the union of a male sperm and a female ovum. Prior to European settlement of the Australian continent, some Aboriginal tribes explained a woman's pregnancy, not by any male–female interaction, but by the invasion of a spirit into the woman's womb. In ancient Egypt, celestial phenomena were explained by reference to several gods, their respective powers and hierarchy. From the time of Isaac Newton, the same celestial phenomena have been explained in terms of deterministic equations of physics.

In modern times though, where superstition and indoctrination are supposed to have given way to reasoned thought, is it still correct to claim that theories (or more particularly scientific theories) are, wholly or partly, socially determined? If social influences are to be totally removed from the processes of theory development, then people cannot have anything to do with these processes. This point is not seriously contested. The question really is: to what extent do social factors affect the content of scientific theories? The effect which a society may have upon the generation of ideas and the development of theories is not usually easy to identify, but neither can it be utterly absent.

At this point of the discussion, we shall digress in order to define the terms: 'Rationalism' and 'Relativism'. The thesis that there is a correct scientific methodology which provides the means of making a rational choice between competing theories is called 'Rationalism'. Those that hold to this thesis are referred to as 'Rationalists'. The conduct of science, to the rationalist is (not surprisingly) eminently rational; scientific research is conducted on the basis of specified principles of rationality. Rationalists claim that the theories of science result from the application of an objective scientific method. Scientific knowledge, to the rationalist, is therefore a unique class of knowledge—it deals with true, existing features of the world. Consequently science is not only superior to other forms of knowledge, it is quite distinct from them. One example of a rationalist philosophy of science is Imre Lakatos's methodology of scientific research programmes.

Rationalists from different schools of philosophy of science do not necessarily agree on a single methodology. Rationalists all tend to agree that there is a correct methodology but each school thinks that its methodology is the correct one. For example, those who advocate the standard view of science would claim that properly performed induction (with its implicit simplicity criterion) is the correct methodology of science; Lakatos would claim that the methodology for scientific research programmes involves choosing progressive research programmes over degenerating ones.

On the other hand, there is the thesis of 'Relativism' or, more correctly, 'Methodological Relativism'. It is rather like the antithesis of rationalism. It states there is no uniquely correct methodology of science. In this view, different methodologies are employed for all sorts of reasons. The primary function of methodology remains the same—to provide criteria for choice between rival theories—but the adoption of one methodology over another is not governed necessarily by any rational factors. Those that hold to this view are called 'Relativists'. Scientific knowledge, to the relativist, is not regarded as being intrinsically different from other forms of knowledge. Thomas Kuhn's theory is one example of a relativist approach to philosophy of science. Another well-known relativist is Paul K. Feyerabend. He argues that the history of science indicates that unchanging rules for the conduct of science do not exist and that instead the only real methodological 'rule' for science is 'anything goes'[4] (see pp. 155–61). Sociologists of knowledge also tend to advocate extreme versions of relativism. However, there exist multiple opinions about the scope and form of cognitive sociology of science; there is no one party line.

THE STRONG PROGRAMME IN THE SOCIOLOGY OF SCIENTIFIC KNOWLEDGE

The 'Strong Programme in the Sociology of Scientific Knowledge' seems to have originated in the Science Studies Unit at the University of Edinburgh in the 1970s. (The strong programme was proposed as a replacement for the previous programme in the sociology of scientific knowledge which made much weaker claims.) The advocates of the strong pr[illegible] the actual content of scientific theories is s[illegible] They claim that such content is greatly affe[illegible]olitical and social structures to the extent that[illegible], laws or processes postulated in scientific th[illegible]ely or at least in large part, determined by t[illegible]least one sociologist goes so far as to suggest t[illegible]t the view that the natural world has *no input* to [illegible]eories!)[5] The advocates of the strong programme do not, ho[illegible]er, all agree on what should constitute the programme.

The definition of knowledge for the purposes of the strong programme will again have to be quite wide in comparison with the traditional philosophical view. One writer on the strong programme is Edinburgh's David Bloor. He defines 'knowledge' for the purposes of sociology of knowledge as follows:

> Instead of defining it [knowledge] as true belief, knowledge for the sociologist is whatever men take to be knowledge. It consists of those beliefs which men confidently hold to live by. In particular the sociologist will be concerned with beliefs which are taken for granted or institutionalised, or invested with authority by groups of men. Of course knowledge must be distinguished from mere belief. This can be done by reserving the word 'knowledge' for what is collectively endorsed, leaving the individual and idiosyncratic to count as mere belief.[6]

Sociologists of knowledge tend to consider their field as a 'science'. This may initially sound quite strange as the methods of sociology would appear, at first sight, to have little or nothing in common with, say, physics. In the analysis of the sociology of knowledge, sociologists take the view that all systems of belief (including science) should be treated as equivalent.[7] Given this approach to knowledge the sociologist makes use of analytical methods appropriate to science in order to 'deconstruct' science itself. On this basis, sociologists have made the claim that the sociology of scientific knowledge encapsulates the values that are found in traditional scientific disciplines. Bloor claims that the values which must be embodied in any version of the strong programme in the sociology of scientific knowledge are found in four principles (or tenets): 'Causality'; 'Impartiality'; 'Symmetry' and 'Reflexivity'.[8] We shall discuss these tenets in due course.

The strong programme in the sociology of scientific knowledge has been subjected to intense criticism from some philosophers. One of the most prominent and vocal opponents of the strong programme is Larry Laudan. He claims that the strong programme is not itself a sociological theory but is instead a meta-theory (or in his words 'a *meta-sociological* manifesto')—a theory which says what broadly should constitute any sociological theory of knowledge:

> [The Strong Programme] must be approached . . . as a set of regulative principles about what sort of theories sociologists should aspire to. Its four constituent 'theses' [tenets] are designed as constraints on the theories which are admissible into sociology. It is important to understand this about the character of the strong programme, since one evaluates regulative principles differently than one evaluates specific theories about social structure and social process.[9]

In regard to the scientific status of the strong programme, Laudan makes some rather crucial points. He notes that not only does Bloor assume that a straightforward demarcation can be made between science and non-science (a very controversial and unresolved issue in itself) but also that Bloor implies that he has found such a distinction. Laudan's evaluation of Bloor's case for the strong programme is well summed up in the following quotation:

> Quite apart from Bloor's specific failure to make the case that the strong programme is 'scientific', the general enterprise seems to put the cart before the horse . . . There is something profoundly paradoxical in saying that we are setting out scientifically to figure out what central features science has.[10]

Bloor has made a strenuous, partially successful counter-strike against Laudan's critique. What's more, the debate (and 'mud-slinging') between rationalist philosophers of science and sociologists of knowledge will, in all probability, continue for some time to come. Bloor's response to Laudan is that one doesn't need to be able to state explicitly all the relevant methods of science in order to do science. Similarly, a scientific approach to examining science itself does not (in Bloor's view) require an explicit knowledge of procedures; these need only be tacit.[11] Yet if the strong programme is indeed a science then some or all of its principles would need to operate in ways similar to procedures found in the natural sciences. We shall now consider in detail Bloor's four tenets of the strong programme.

The tenet of 'Causality' asserts that all states of belief are caused by and within a given social context, that is, that beliefs have social causes (amongst others). It is not sufficient to justify a particular belief merely by saying that it is rational to hold this belief or that this belief is true. Philosophical justifications for holding a belief have traditionally been of this form—a belief follows from accepted premises by a rational process of inference. On the traditional philosophical account, no further justification for holding a belief is necessary. Advocates of the strong programme deny that such justification is sufficient because all beliefs are to some extent socially determined and this aspect of belief formation is in need of explanation.

An example regularly cited is the case of the initial development of quantum mechanics. Paul Forman has argued in a well-known article that the social conditions in post-World War I Germany led to the abandonment of the physical principle of causality (the principle that all events in the natural world have a cause). The defeat of Germany and its post-war conditions led German public opinion to an anti-intellectual and anti-science stance. The principle of causality was, according to Forman, singled out as representative of those doctrines which were no longer acceptable. The world-view of German physicists was correspondingly modified from that which was held before World War I. This change in world-view was prior to the establishment of quantum mechanics. Unlike the theories of classical physics, quantum mechanics does not require all physical events to be caused. In this German context the theory of quantum mechanics is relevant since, had it emerged earlier, quantum mechanics could have been used by the German scientific community to justify the abandonment of the principle of causality. Forman argues that the opposite situation occurred—the intrinsically a-causal nature of quantum mechanics appears to have been an outcome of prevailing social conditions.[12]

Laudan describes the tenet of causality as being 'a regulative principle' which is 'relatively unproblematic'. It is unproblematic because the implication that true or rational or successful beliefs are not in need of any further explanation, other than that they are true or rational or successful, is not the same as asserting that they are uncaused or have no social causes.[13] If all beliefs are caused, this in no way entails that all or even most causes are necessarily sociological. Mary Hesse of Cambridge offers the example of someone performing simple arithmetic.[14] This person does the arithmetic operations in a particular way because of social pressure, for example, the teacher says to do it this way and not any other. Hesse claims that people do not do these arithmetic operations in the manner prescribed because it is logical to do so. Hesse is correct in saying that there is a social cause for the action of doing the arithmetic by the usual rules. This, however, has no bearing on whether the arithmetic procedures are themselves logically correct, nor does it exclude there being other reasons for doing these operations as prescribed.

In regard to the claimed scientific status of the strong programme, the tenet of causality does not assist in establishing sociology as a science, for it is not even the case that all scientific theories require causal explanation—for example, in quantum mechanics.[15]

The tenet of 'Impartiality' asserts that all beliefs, be they true or false, rational or irrational, successful or unsuccessful, require explanation. Explanation should be impartial to the form of the belief. Older schools of sociology of knowledge tended to accept that only flawed beliefs were in need of explanation. The type of argument used goes something like this: objective scientific methods produce certain knowledge (that is, certain in the sense used by non-cognitive sociology of science). If, however, one discovers that some scientific belief is incorrect then there must have been other factors external to science (for example, social ones) that caused this belief to be accepted. Only false or irrational or unsuccessful beliefs are, in this sense, *caused*. Thus the sociology of scientific knowledge had originally been restricted to the sociology of error. Bloor questions the notion that causation should only be associated with error. He argues that it would be a more plausible situation if causes brought about all kinds of beliefs.[16]

Laudan argues that this tenet is redundant since, if all beliefs have causal explanations as required by the tenet of causality, then this regardless of whether the beliefs are true, false, rational, and so on. As this tenet follows from the tenet of causality, its status as scientific is as open as that of the tenet of causality.[17]

The 'Reflexivity' tenet arises from accepting sociology of knowledge as a science. If the sociology of scientific knowledge is itself a science then it must also be subject to the same analytic criteria as it sets out for explanations of the natural sciences.[18] The reflexivity tenet also follows from the tenet of causality (which refers to all beliefs) and is therefore also redundant. As a consequence, the reflexivity tenet can shed no light on the scientific status of the strong programme.[19]

The tenet of 'Symmetry' asserts that the same type of cause can bring about both true and false, or rational and irrational, or successful and unsuccessful beliefs. One does not have to postulate, for example, one type of cause exclusively to explain true beliefs and another exclusively to explain false beliefs. Both kinds of beliefs in a given instance of scientific knowledge can result from the same cause. Suppose that a researcher decides to investigate lung cancer rather than some other disease because there is readily available funding for this research. The researcher then makes substantial findings which change our understanding of cancer development. Further suppose that it is later discovered that some (but only some) of this researcher's conclusions were incorrect. The sociologist of knowledge might attempt to explain this situation of the generation of both correct and incorrect beliefs by appeal to the one cause, for example, the pressure on the researcher to gain adequate funding.

Laudan splits the tenet of 'Symmetry' into three parts which he labels 'Epistemic', 'Rational' and 'Pragmatic'. We shall presently consider each of these individually. Laudan also says that, in general, the symmetry principle goes against well-established precedents of the natural sciences.[20] In the natural sciences (for example, in physics or chemistry) different causes are invoked to explain different phenomena. This question about whether the same type of cause can produce different beliefs is one, says Laudan, that can only be settled by experiment. If this were done and the result turned out to support the symmetry thesis, then this thesis could lay claim to a scientific status. Without such experimental evidence in its favour, Laudan claims that the tenet of symmetry merely offers an *a priori* answer to a question that can only be settled empirically.[21]

We shall now consider the different parts into which Laudan splits the symmetry tenet. 'Epistemic Symmetry' is an assertion to the effect that the same type of cause can bring about both true and false beliefs. Laudan's attitude to the truth-status of scientific beliefs is that we cannot know absolutely if these beliefs are true or false. In other words, we do not have epistemic access to the truth-status of our theories.[22] If we cannot know absolutely that our theories are either true or false then we also cannot know whether the symmetry thesis applies or not. The question of epistemic symmetry therefore evaporates.

'Rational Symmetry' is an assertion to the effect that the same type of cause can bring about both rational and irrational beliefs. Without going into the very wide and complex nature of theories of rationality, it will suffice to define basically a causal theory of rationality.[23] Following Laudan, we state that a causal theory of rationality is one in which a rational individual forms a belief on the basis of reasons which are related to the individual's aims and knowledge.[24] Reasons are thus causes of states of belief. On the other hand, if the tenet of rational symmetry is accepted then any assessment of a belief becomes irrelevant to explaining its formation and human reasoning would play no causal role in the generation of beliefs. Laudan describes such a situation as *absurd.* William Newton-Smith of Oxford comments that the thesis of rational symmetry would destroy the distinction between propaganda and rational argument. He further adds that if this is the view of the Edinburgh School, then they might as well attempt to bribe everybody else into agreeing with them![25] The Australian philosopher, F. John Clendinnen, adds force to this by pointing out that there is a genuine source of paradox in Bloor's argument:

> If it is held (as Bloor seems to) that all there is to a belief being called rational is that it conforms to certain socially accepted criteria, then what can be the point in persisting with any beliefs once this is recognized?[26]

'Pragmatic Symmetry' asserts that the same type of cause explains both successful and unsuccessful beliefs. Laudan notes that empirically successful theories tend to be accepted for longer periods of time than unsuccessful ones. This implies that the success or otherwise of a theory does have a bearing on how long the theory will be accepted.[27] Bloor's response has been to clarify his position in relation to pragmatic symmetry and its implications:

> The pragmatic success of a theory is, I am sure, often connected with its acceptance and espousal by the scientific community . . . Even the acknowledgement that a theory *is* pragmatically successful involves complex judgements and is frequently a matter of dispute. Success here has to be weighed against failure elsewhere. The past history and future prospects of a theory have to be judged and compared with rivals. When pragmatic success is put in context its indications are never equivocal nor as simple as they may seem in the abstract.[28]

This would have been a suitable point at which to depart from discussion of the strong programme if it were not for developments in the rapidly expanding field of Cognitive Science studies, especially Artificial Intelligence (A.I.).[29] Results from A.I. research have been used to argue against the strong programme. In particular, articles by Peter Slezak have generated much debate.[30] Slezak claims that the strong programme is refuted because computers can now make discoveries of classical scientific laws. Slezak is referring to the programming of very sophisticated computers with software (computer programs) called 'BACON' and 'AM'.[31] The BACON program has allowed computers to generate laws of physics (for example, Kepler's Third Law of planetary motion) and of chemistry that were originally developed in very different eras and social settings. Aside from its programming, all the computer requires is an input of raw empirical data in order to produce physical laws. Slezak argues that the production of scientific laws by computer is independent of any social situation and therefore refutes the claim that scientific knowledge is socially constructed. He writes:

> The success of BACON and AM programs provide empirical confirmation of the view that the specific historical circumstances which undoubtedly attend scientific reasoning do not play a decisive role . . . On the contrary, the programs demonstrate that the widely varying social factors attending the original discoveries played no part in determining their specific contents.[32]

The responses to Slezak's paper have come thick and fast. This was, of course, highly predictable since there are very large professional and personal stakes involved. Things endangered when whole programmes of research come under threat (in this case the strong programme) include: cherished and steadfastly held positions on the issues involved; the academic prestige of the discipline; the respected opinions of one's peers and one's individual status as a researcher. The manner and tone of the debate surrounding Slezak's claims can been seen from different perspectives. On one rendering, it may be viewed as an attempt to finally put the sociologists back in their rightful place after years of their invading the proper enquiry ground of philosophical and cognitive studies of science. Alternatively, the situation can be viewed as an effort by the cognitive science investigators to achieve dominance in the study of science by staging a cognitive coup. Such a coup will not only show that philosophers of science are mostly impotent on cognitive issues and that the sociologists of knowledge are wrong, but will also show that the sociologists don't really know what they are talking about! Other interpretations of the rhetoric are also possible. Let's now look at the principal objections to Slezak's conclusion.

The sociologist Harry Collins claims that Slezak's premise that BACON makes scientific discoveries is false, in which case Slezak's conclusion regarding the strong programme is also false. Collins says that social influences do appear in a number of ways. Human beings choose the data to be fed into the computer and

Collins infers that such sifting of data predetermines the computer result. He denies that computers make scientific discoveries because the operations performed by the computer are only part of what we might call 'the process of discovery':

> For most individuals, however isolated, there is more to 'discovery' than data-trawling and statistical induction. The 'more' is the ability to anticipate the social nexus into which the potential discovery will be cast . . . To do this the potential 'discoverer' has to be in touch with social life . . . The calculus of risk and ridicule, the calculus of possibility, depends not on what relationships can be extracted from the data, but on how different interpretations of the numbers will be received.[33]

Ron Giere argues on a slightly different tack. He claims that the interests and the expectations of those who have designed the programming heuristics of BACON (or similar programs) are contained within the software. He justifies this claim by noting that the general mathematical structure of the physical laws to be searched for by BACON are prescribed by the program creators and embodies their interests.[34] The content of the laws 'discovered' by the computer would, therefore, be affected. In other words, there would still remain some social causation.

Paul Thagard's comments are much more sympathetic to Slezak's argument but he does not accept that the strong programme is refuted. Rather Thagard believes that the cognitive and sociological approaches are complementary.[35]

Slezak has responded in kind to these and other criticisms. In response to Collins, Slezak accuses him of failing to understand two important features of A.I. Firstly, Slezak denies that the data input to BACON predetermines the outcome and that Collins has confused the issues of choosing the data from making a choice consistent with the data:

> Collins assumes that the results generated by BACON must be predetermined by the selection of the data provided to it through having been 'filtered' by the 'human social collective' in accordance with prevailing thinking. But in that case there is no sense in which there is a choice among alternative generalizations at all—contrary to his own supposition . . . certainly the data we give to BACON will constrain

the outcome—just as it did for Kepler—but it does not determine which among alternative possibilities will be inferred, if any.[36]

Collins's second confusion, according to Slezak, is that he appears to think that purely algorithmic methods (such as are found in pocket calculators) are no different in principle to the heuristic problem-solving methods used in A.I. Slezak claims that this distinction is fundamental to the study of A.I.[37]

Slezak agrees with Giere that it is impossible to devise a program that can search all possible solutions to a given problem. However, Slezak points out that if the interests of the designers are embodied in the computer programs, it is only in the sense that the program contains an 'interest' in solving the problem by more efficient means than by 'blind' search.[38]

One further point to note, also made by Thagard, is that BACON and AM are only the beginning of these sorts of computer programs and what might be achieved with such programs cannot be predicted. The arguments offered by the proponents of the strong programme may have averted disaster for the time being, but the impact of future A.I. studies looks potentially very damaging for the strong programme.[39]

THE ANALYSIS OF SCIENCE IN TERMS OF INTERESTS

The strong programme in the sociology of scientific knowledge is really a set of guiding principles to which sociological theories of scientific knowledge should conform. These principles are designed to indicate the methods and procedures appropriate for such sociological theories, but the strong programme is not itself a sociological theory. There are several theories of sociology of scientific knowledge that conform to the guidelines of the strong programme. One such account of cognitive sociology of science which has received much attention is the 'Interests Analysis'. Advocates of the strong programme (such as Bloor and Barnes) explicitly and enthusiastically embrace the interests analysis of science. Sociological theories of scientific knowledge did not, of course, start with the advent of the strong programme. For example, Marxist theorists have tended to interpret the creation and content of scientific theories in western capitalist countries as determined by the dictates of the ruling class. Scientific theories in this context are 'instruments' for the purposes of repressing the lower classes. Much recent interests analysis also follows from the writings of Jürgen Habermas.[40]

The analysis of science in terms of interests is claimed by its advocates to be a naturalistic approach to the subject matter. What can they be asserting by this? A general definition of 'Naturalistic Inquiry', or 'Naturalism', is: 'the view that all human activities are to be understood as entirely natural phenomena, as are the activities of chemicals or animals'.[41] Yet those engaged in interests analyses either don't agree on the definition of naturalism or simply don't define it at all. Senses of naturalism range from it meaning a form of enquiry similar to that of the natural sciences to a descriptive account rather than a prescriptive one.[42] This lack of agreement on an explicit definition for naturalism is a possible source of confusion since explanations utilizing interests may depend on different renderings of naturalism.

In interests analyses, scientific theories are viewed as 'resources' or 'instruments'. Scientific theories are one means by which a group of scientists can further their interests. What might these interests be? Initially we shall distinguish two broad categories of interests—'Scientific' (or 'Instrumental') interests and 'Social' interests. Le Grand offers some examples of these:

> Scientific interests might include one's expertise in particular techniques or one's greater skill in theorization as opposed to experimentation. Social interests range from narrower ones which might include religious or political or other social beliefs or one's economic or social position in the larger society.[43]

Barnes argues that the production of knowledge is facilitated under an open (publicly acknowledged) interest in the prediction, manipulation and control of nature. He further contends that knowledge grows as a result of this interest and also by a concealed (covert) interest in rationalization and persuasion.[44] The interest in prediction, manipulation and control is a scientific one. We want our scientific theories to make predictions about future events and future states of physical systems. Accurate prediction allows for detailed planning of future action. An ability to influence physical systems offers the possibility that we may be able to direct them in ways which we find desirable. Our scientific theories tell us how we can affect the development of physical systems. Control comes about through a combination of the abilities to predict and to manipulate. Scientific theories can, therefore, assist in establishing limited control over some natural phenomena. The exact form that the interest in prediction, manipulation and control may take is context, or culturally, dependent. For example, in a purely agricultural-based society, an ability to predict adverse weather conditions may allow for action that would help ensure crop survival.

The interest in the prediction, manipulation and control is not considered by sociologists as sufficient to determine fully scientific knowledge, so other interests need to be brought in to complete the explanation.[45] The interest in rationalization and persuasion is a social one. In scientific communities there is a strong desire to conduct affairs in a sensible and rational manner. After all, scientific research is at least perceived to be dictated by logical and rational considerations. The high status of science as an institution and of the scientists themselves is dependent on science being perceived in this manner. (Recall the discussion at the end of Chapter 5 on the *post hoc* explanations constructed by scientists in order to give the appearance of a rational process.) Other consequences important to any scientific community also follow from the appearance of rationality: for example, reinforcement of professional ideals and achievements; levels of funding and so on. Thus there is a very great social interest in scientific communities to ensure that at least the appearance of science as a rational exercise is maintained, regardless of what actual scientific practices are currently employed. This interest in the maintenance of the appearance of rationality is not publicly recognized or explicitly stated, for to do so would be to acknowledge that the scientific endeavour need not be conducted in a manner consistent with rational norms.

Barnes claims that when knowledge is created or accepted, the particular form of this knowledge is due to social interests. Where the interests are not open but instead are hidden, Barnes says that the beliefs generated are ideologically determined:

> whenever knowledge is ideologically determined there is a disguise or concealment of an interest which generates or sustains the knowledge ... Knowledge or culture is ideologically determined in so far as it is created, accepted or sustained by concealed, unacknowledged, illegitimate interests.[46]

A frequently cited example is the case of statistics in the late nineteenth and early twentieth centuries. The development of statistical theory in Britain during this time and its relation to prevailing interests has been a major source of study. Donald MacKenzie has made a well-documented study of this case utilizing interests analysis.[47]

Statistics is usually considered as a branch of applied mathematics. This being so, one would expect that the content of statistical theory ought only to depend on its initial definitions, theoretical premises and the rules of deductive inference. MacKenzie argues that the interests of the researchers who developed statistical theory (Galton, Pearson and Fisher) affected the actual theoretical content of their work! These researchers all had a strong interest in eugenics—the study of the production of better human offspring. One use of statistical theory is the prediction of the characteristics of future populations. We have already noted that accurate prediction offers the opportunity to exert some influence on the likely course future events might take. Thus successful statistical techniques could be utilized as an instrument for prediction and control. This is not the whole story though. The explicit interest in eugenics by Galton, Pearson and Fisher was one specific form of more general social interests, namely, the social interests of the rising British middle class.[48] It may therefore be argued that the development of statistical theory in Britain was ideological determined, in the sense of Barnes's definition. MacKenzie's conclusion from this case-study is stated in the following extract from his book:

> Science is an activity not of passive contemplation and 'discovery' but of invention. It is goal-oriented, and, while its goals may all in a general sense have to do with the enhancement of the human potential to predict and control the world, they represent different particularisations of this overall objective. The pursuit of particular goals is typically sustained by social interests located either in the internal

> social structure of science or in that of society at large. Scientific knowledge is thus a social construct in two senses. First, in that it is typically the product of interacting groups of scientists. Second, in that social interests affect it not merely at the organisational level but at the most basic level of the development and evaluation of theories and techniques. Because science is goal-oriented, and because its goals are socially sustained, scientific knowledge is a social construct.[49]

It has been further argued by Barnes and MacKenzie that instrumental interests 'pre-structure' (or pre-organize) a scientist's choice of theory.[50] In other words, theory choice is influenced by the instrumental interests of a scientist (or more appropriately, a specialist group of scientists). The instrumental interests are themselves related to sets of social interests, for social interests may determine what sort of prediction is important, or what should be considered a problem.[51] What was not considered a problem yesterday may well be an important problem tomorrow! These social interests need not be exclusively restricted to a scientific community. They may also be concerns of the general community. In MacKenzie's case-study, for example, he argued that the content of statistical theory was influenced by an interest in eugenics. This interest was not restricted to scientists as it had a basis in the wider society of the time, especially within the professional British middle class.[52] The case can therefore be made that instrumental interests provide the scientist with a certain predisposition towards choosing a particular theory and that these interests may be due to, or reinforced by, existing social interests.

Whilst sociologists have raised pertinent points about the social environment in which scientific theories develop and the effects of such environments, the cogency of their arguments and the validity of their conclusions have been seriously challenged. Compare, for instance, MacKenzie's above conclusion regarding the efficacy of social interests with that of Slezak:

> Demonstrating the usefulness of some theory as a vehicle for promoting social interests, or even its role in determining patterns of acceptance and rejection, does not *ipso facto* establish those factors as causes

> of the theory's contents. Indeed, far from being 'beyond dispute', as Bloor has asserted, the causal claims are sufficiently implausible to be *beyond belief*. . .[53]

Interest analysis does provide useful insight into theory choice and this will be pursued further in Chapter 7.

FEYERABEND ON SCIENCE AND MYTH

Paul K. Feyerabend (1924–) was born in Vienna and gained his doctorate from the University of Vienna. He has explicitly described his position on scientific method as 'anarchist' and argues strongly against the rationality of science and the superiority of scientific knowledge. Feyerabend's philosophical razor has cut deeply into both the traditional and recent philosophy of science. Despite his probing critiques of many cherished arguments of other philosophers of science, Feyerabend's own theory about the nature of scientific theories belongs more to the arena of the sociology of scientific knowledge than to mainstream philosophy of science. Feyerabend's ideas on the institution of science, its relation to society and scientific methodology are found in his books *Science in a Free Society*, *Against Method* and *Farewell to Reason*. Feyerabend holds appointments as a Professor of Philosophy at the University of California (Berkeley) and as Professor of Philosophy of Science at the Federal Institute of Technology (Zurich). During his academic career he has also lectured at the universities of Bristol, Sussex, Auckland, Yale, the Free University of Berlin and at University College, London.

Feyerabend claims that the history of science is far richer than would be possible under fixed rules and that all attempts to describe science using fixed rules find severe difficulties when historical evidence is brought to bear. He argues that the history of science is so complex and varied that no single methodology of science could ever be uniquely applicable, that is, all rational rules for the development of science have been violated at some stage. He writes:

> the idea of a fixed method, or a fixed theory of rationality, rests on too naive a view of man and his social surroundings . . . there is only *one*

> principle that can be defended under *all* circumstances and in *all* stages of human development. It is the principle: *anything goes*.[54]

Feyerabend cites a number of historical cases to support his argument. He gives most attention to the development of astronomy and physics from the time of Aristotle to Galileo. Feyerabend claims that during this development there were instances where rational choices were not made due to the dominance of what he calls 'irrational elements'.[55] These irrational elements include such things as being obstinate or arrogant, using propaganda, *ad hoc* adjustments and even the making of errors. In Feyerabend's view, we would not hold to some so-called rational beliefs today, such as a heliostatic solar system, if it were not for the effects of these irrational elements. Galileo's championing of the Copernican theory was, according to Feyerabend, done by breaking 'rational' rules and by Galileo becoming a propagandist for Copernicanism.

One powerful argument at the time of Galileo against the Copernican theory was that if an object was dropped from the top of a tower, it would land at the base of the tower and not some distance away. This was alleged to show that the Earth was indeed motionless. If the Earth rotated as suggested by Copernicus, then in the time required for the object to fall to the ground, the base of the tower would have moved by many metres. When this experiment was performed and the object landed at the base of the tower, it was taken as an empirical refutation of the Copernican theory.

What does Feyerabend contend about the way in which Galileo proceeded? Feyerabend claims that Galileo makes a new interpretation of empirical results (an object falling to a tower's base), one which is in conformity with Copernican theory. What Galileo achieved by such a move as this is to introduce another observation language.[56] (Recall from Chapter 1 that an observation language is defined in terms of some theory, rather than being theoretically independent.) Galileo does not do this openly, says Feyerabend, instead he disguises the new interpretation of experience with his rhetoric:

> Galileo's utterances are indeed arguments in appearance only. For

> Galileo uses *propaganda*. He uses *psychological tricks* in addition to whatever intellectual reasons he has to offer . . . But they obscure the new attitude towards experience that he is making . . . They obscure the fact that the experience on which Galileo wants to base the Copernican view is nothing but the result of his own fertile imagination . . . by insinuating that the new results which emerge are known and conceded by all, and need only be called to our attention to appear as the most obvious expression of the truth.[57]

By adhering to the Copernican theory, Galileo was holding to a theory that was inconsistent with the common sense interpretation of the tower experiment and several more besides.[58] It is further argued by Worrall that, by adopting the Copernican theory, Galileo embraced a theory that was inconsistent with other well-confirmed theories at that time, especially Aristotelian dynamics, which states that the Earth is motionless. Galileo, therefore, could be accused of violating two rational principles: the principle that a theory incompatible with experiment should be abandoned and the principle that a theory should not be held if it is inconsistent with currently accepted ones.[59] In addition, Feyerabend claims that Galileo makes use of *ad hoc* adjustments in order to overcome initial objections to the Earth's motion through space and to allow 'breathing space' for the Copernican theory to develop.[60]

Feyerabend generalizes these (and other) considerations in a manner such that he can claim that the history of science shows any methodological rule can be dispensed with, if the researcher so desires and for all kinds of psychological, social, personal or political reasons. This conclusion is, however, not justified. Leaving aside the objection that Feyerabend's general conclusion—the non-existence of proper scientific methodology—is inductively inferred, it would appear that the mere suspension of one or more methodological rules may well be in accord with other aspects of prevailing methodology. Rules may not be adhered to for a period of time in order to allow for further development in the face of problems confronting the theory, problems which might otherwise halt any form of progress. This has been suggested by several commentators, notably Kleiner, who explicitly states:

> [Feyerabend's] argument from the history of science, is blatantly fallacious . . . From the proposition that every rule has been justifiably violated on some occasion, it simply does not follow that no rule is ever in force. The premises do not rule out the possibility that other methodological rules, rules based on other methodological desiderata, remain in force on the occasion cited.[61]

It would be easier for Feyerabend to substantiate a less radical thesis on scientific methodology than his 'anything goes'. What emerges from the history of science is that although there is no absolute scientific method and accompanying scientific methodology that is employed by all scientists at all times, there are nevertheless methods used. What specific methodology applies seems to depend on a number of factors, not the least being the stage of development of a particular science. What is an appropriate methodology at one time may well be inappropriate at another time when different theoretical assumptions are accepted.

Feyerabend also argues that theoretical pluralism—the existence of more than one theory in a given field at any one time—is the norm in science.[62] Theoretical pluralism is not only consistent with Feyerabend's 'methodological rule' of 'anything goes', it is well in keeping with the spirit of this 'rule'. If a researcher is free to pursue any line of investigation, then a multiplicity of theories should be expected. Again, he cites cases in the history of science to substantiate his argument and it seems fairly clear from our earlier chapters that Feyerabend is correct on this particular point. Feyerabend thinks that such proliferation of theories and the resultant competition between them is the real source of progress in science.[63]

On the topic of the incommensurability of scientific theories, Feyerabend's writings predate Kuhn's. Feyerabend accepts that the grammar of any natural language, such as English, Greek or Mandarin, incorporates a specific view of the world.[64] The world-view implicit in a language is contained both in the meanings of the individual words and in what is called 'covert classifications'. These classifications determine which words belong to particular classes. For example, Christian (or given) names are classified by sex (male or female) even though the names themselves need not be associated with one or other sex. The world-view contained within a language influences both the perceptions and the thought patterns of the people who speak and think in that language. Feyerabend argues that covert classifications act as sources of psychological resistance to alternative world-views. Since incommensurability depends on covert classifications, Feyerabend says that it is difficult to explicitly define, but he does offer the following description: 'If these resistances [to alternative world-views] oppose not just the truth of the resisted alternatives but the presumption that an alternative has been presented, then we have an instance of incommensurability'.[65]

Feyerabend argues that high-level theories can be considered in a manner akin to natural languages, in that the terms of the theories have both overt meanings and covert classifications. Thus incommensurability between scientific theories is to be expected.[66] Feyerabend's concept of the incommensurability between different scientific theories is similar to Kuhn's—the contents of the different theories cannot be directly compared because of a logical incompatibility.[67]

Whether there is incommensurability between theories will also depend on what interpretation is placed on these theories. Suppose, for instance, the terms of the theories are not interpreted literally and the theories are merely considered as 'instruments' for making calculations and predictions (that is, given an instrumental interpretation). Feyerabend would then say that the theories will be commensurable if they can be related to the same observation language:

> the question 'are two particular comprehensive theories . . . incommensurable?' is not a complete question. Theories can be interpreted in different ways. They will be commensurable in some interpretations, incommensurable in others. Instrumentalism, for example, makes commensurable all those theories which are related to the same observation language and are interpreted on its basis.[68]

Given that Feyerabend claims anything goes in science and that scientific theories are incommensurable, where does he stand on the issue of the status of scientific knowledge? Rationalist philosophers of science argue that there is a strong case for the objectivity of scientific knowledge. One reason usually offered why science should have a special status is because it is grounded in the natural world. Feyerabend has argued that science is an ideology—one belief system amongst many others—and has no more claim to a higher status than does any other belief system. In the category of belief systems Feyerabend explicitly includes myths, voodoo and witchcraft! He claims that such belief systems as these not only have traits in common with science, they are just as well based in the natural world as science is.

It is generally conceded that there is some factual basis, however small, to all myths but Feyerabend further expresses the view that many myths are highly developed systems of belief based on observations. This would require myths to have an empirical basis. If this is the case, then science itself cannot be of much assistance in separating myth from science and one might expect there to be some overlap between myth and science.[69] In what ways might this occur? Feyerabend claims that both begin with experience—that is, observations—and then they are developed theoretically. Feyerabend thinks that at a fundamental level there is no difference between scientific observations and those attributable to myths.[70]

What about the theoretical development of science? Is this not different to that found in non-science, and especially in myths? But no, Feyerabend claims that the theoretical development of both science and myth runs parallel; the making of any theory, scientific or otherwise, follows the same general path:

> both science and myth cap common sense with a theoretical superstructure ... Theory construction consists in breaking up objects of common sense and in reuniting the elements in a different way. Theoretical models start from analogy but they gradually move away from the pattern on which the analogy is based. And so on. These features ... refute the assumption that science and myth obey different principles of formation ... that myth proceeds without reflection ... or speculation.[71]

In drawing a comparison between science and other belief systems, Feyerabend claims to have shown that neither one is superior to the other. Whatever else science may be, it is correctly characterized as a belief system because people believe the claims made by science. What Feyerabend has done is to identify traits that are common to all belief systems. This, however, does not imply any equivalence amongst such systems, nor does this by itself say anything much about the status of the beliefs generated by any sophisticated belief system. John Worrall points out that, although Feyerabend indicates that science need not necessarily be any more objective in revealing facts about the world than are voodoo or witchcraft, he fails to say how it is that witchcraft as a theory explains the world.[72] This is a most important criticism of Feyerabend's argument about the nature of scientific knowledge. Scientific theories both qualitatively explain and quantitatively predict to an amazing degree of accuracy. A distinction can, *ipso facto*, be made between science as a system of beliefs and other such systems on this very basis.

LATOUR ON THE SOCIAL CONSTRUCTION OF SCIENTIFIC FACTS

Bruno Latour's academic background is in the field of social anthropology. In more recent times he has applied methods of investigation traditionally used by anthropologists to the study of scientific research. In pursuit of a social anthropological analysis of science, Latour spent two years working with laboratory scientists at the Salk Institute for Biological Studies. Latour's books, *Laboratory Life* (co-authored by Steve Woolgar) and *Science in Action*, are an outcome of his experiences at the Salk Institute. Bruno Latour holds a position as an Associate Professor of Sociology at the Centre de Sociologie de L'Innovation of the Ecole Nationale Supérieure des Mines in Paris, France.

Latour believes that what we call scientific facts are not discovered. Instead he argues that they are socially constructed; they arise as a result of complex social processes. He further claims that 'Nature' is not what exists but what results when a controversy in science is settled. Latour is perhaps the most prominent of those who espouse this 'Constructivist' view of science. He thinks that science as a whole or even at the level of the individual discipline is such an enormous enterprise that a thorough sociological analysis cannot be done, due to the sheer size of such an undertaking. The only way for the anthropological sociologist to proceed is by descending to the grass roots of science; the sociologist must get into individual laboratories and must look at individual scientific articles. This is the approach Latour adopts. He takes the reader on a convoluted, imaginary journey through scientific journals and laboratories with himself as guide, narrator (and bottle-washer). We shall follow Latour to see what he has to offer.

Our journey begins with Latour defining what he calls 'black boxes':

> The word [sic] **black box** is used by cyberneticians whenever a piece of machinery or a set of commands is too complex. In its place they draw a little box about which they had to know nothing but its input and output . . . no matter how controversial their history, how complex their inner workings, how large the commercial or academic networks that hold them in place, only their input and output count.[73]

He applies this term to anything in science which is uncontroversial—anything whose origin and internal workings are not questioned but taken as known in the normal routine of scientific research. Examples of black boxes include computers, spectrometers, DNA, and so on. This notion of black boxes is important to Latour for he claims that controversies end when black boxes are closed and when one opens a black box, one creates a controversy.

A related concept is the 'modality'. Latour defines modalities as those sentences that qualify (modify) other sentences.[74] There are both positive and negative modalities. A positive modality changes a statement in such a way that the origin of the statement is less obvious to someone reading or hearing the sentence. A positive modality qualifies a statement such that its new, modified form may entail consequences not previously entailed in the original form. In contrast, negative modalities change a statement such that its origin is far more obvious. Negative modalities also indicate why the statement was considered as being strong or firm, flimsy or unfounded. Latour writes:

> We shall call **positive modalities** those sentences that lead a statement away from its conditions of production, making it solid enough to

render some other consequences necessary. We will call **negative modalities** those sentences that lead a statement in the other direction towards its conditions of production and that explain in detail why it is solid or weak instead of using it to render some other consequences more necessary.[75]

Latour provides some explicit examples of these modalities in action. Let's consider one of his more serious examples:

(5) The primary structure of Growth Hormone Releasing Hormone (GHRH) is Val-His-Leu-Ser-Ala-Glu-Lys-Glu-Ala.

(6) Now that Dr Schally has discovered . . . (the primary structure of GHRH), it is possible to start clinical studies in hospital to treat certain cases of dwarfism since GHRH should trigger the Growth Hormone they lack.

(7) Dr A. Schally has claimed for several years that . . . (the structure of GHRH was Val-His-Leu-Ser-Ala-Glu-Lys-Glu-Ala). However, by troubling coincidence this structure is also that of haemoglobin, a common component of blood and a frequent contaminant of purified brain extract if handled by incompetent investigators.

Sentence (5) is devoid of ownership, construction, time and place . . . It is, as we say, a **fact** . . . it is inserted into other statements without formal modification . . . sentence (5) becomes a closed file, an indisputable assertion, a black box.

Latour's sentence (6) is a positive modality since, in his view, anyone believing it could be led to use the GHRH in further studies without questioning its structure. Latour's sentence (7) is a negative modality since anyone believing it is presented with an open question about whether such a hormone was developed at a certain New Orleans laboratory or not. Latour adds:

According to which direction we go, the original sentence (5) will change status: it will either be a black box or a fierce controversy . . . Inserted inside statement (6), (5) will provide the firm ground to do something else; but the same sentence broken down inside (7) will be one more empty claim from which nothing can be concluded.[76]

Starting with the quotation immediately above, we first draw attention to the fact that sentences do not change their status; they remain sentences. What Latour appears to be saying here is that propositions expressed by sentences may be given a different interpretation depending on what other propositions are concurrently asserted. What's more, if one sentence is *inserted* into another, what results is a *new* sentence with perhaps a different meaning. So when Latour cuts up his sentence number (5) and places other phrases around it, he makes new sentences which assert something different to what the original sentence did. (People have been doing this for years!) What's the point of the whole exercise?

The point in question is that Latour wants to demonstrate (in a novel way) how the construction of a sentence can eventually affect the beliefs which people will hold. In particular, he wants to show how some forms of rhetoric can be more persuasive than others when they involve assertions which are presented as firmly established or undisputable—that is, as black boxes. This is the beginning of fact construction, according to Latour:

> A sentence may be made more of a fact or more of an artifact depending on how it is inserted into other sentences. *By itself a given sentence is neither a fact nor a fiction*... You make it more of a fact if you insert it as a closed, obvious, firm and packaged premise leading to some other less closed, less obvious, less firm and less united consequence.[77]

Now, of course, sentences are neither facts nor artifacts in themselves. The proposition asserted by a sentence may express a fact or it may not. It is, however, more correct to say that the proposition expressed embodies what *we take* to be factual and such statements are always subject to challenge and possible revision. There isn't anything essentially novel in stating this though. One would not be unjustified in describing this discourse of Latour's as nothing more than an exercise in reinventing the wheel and providing said wheel with an alternative name!

The next phase of Latour's process of fact construction involves not just the individual researcher, but many. Latour claims that what others do with a statement after it has been proposed will determine whether the statement becomes a fact or a fiction. If a statement is added to other phrases or sentences so as to result in a positive modality and this process is continued, the original statement will become a fact. If, on the other hand, what results is a negative modality and that process continues, the original statement becomes a fiction. Latour writes:

> the fate of what we say and make is in later users' hands . . . By looking at them (i.e. statements, machines) and at their internal properties, you cannot decide if they are true or false, efficient or wasteful, costly or cheap, strong or frail. These characteristics are only gained through *incorporation* into other statements, processes and pieces of machinery. These incorporations are decided by each of us, constantly. Confronted with a black box, we take a series of decisions. Do we take it up? Do we reject it? Do we reopen it? Do we let it drop through lack of interest? Do we make it more solid by grasping it without any further discussion? Do we transform it beyond recognition? This is what happens to others' statements, in our hands, and what happens to *our* statements in others' hands. To sum up, the construction of facts and machines is a *collective* process.[78]

The questions posed by Latour in the above quotation bear a striking similarity to the questions asked when attempting to decide between competing theories. It is interesting to note that although Latour says that we ask these sort of questions and that we constantly make decisions about them, he never actually spells out how these decisions are arrived at.[79]

When a researcher claims to have made a factual discovery, this will not usually be accepted by all other researchers in the relevant field and a controversy may result. The 'discoverer' of the fact will want to convert it into a closed black box and if he or she encounters opposition, Latour says that a 'battle' between the discoverer and his or her opponents begins in the scientific literature. What is more, as this battle intensifies, the content of the scientific articles written on the subject becomes highly technical.[80] Latour uses the term 'technical' to include the opening of more and more black boxes so that the controversy between researchers covers not just the initial 'discovery', but a growing number of disputed issues. More questions are asked, more facts are scrutinized, more black boxes are opened and the 'war of paper' continues to expand.

An isolated researcher who 'fights' alone is doomed, according to Latour. There will come a stage in the controversy when the resources available to the individual will be insufficient to defend the position taken. Additional resources must be found and a ready source is the published work of other researchers. Latour thinks that by citing many references, a researcher is actually fortifying his or her own position and by doing so, may be making the crucial step in transforming an opinion into a fact. The reason—if a scientist attacks the conclusion of a scientific paper then he or she also attacks all the literature cited in that article. If the number of citations is large, then this can become a very tedious and daunting task. Latour strongly emphasizes the effect of including numerous citations:

> The presence or the absence of references, quotations and footnotes is so much a sign that a document is serious or not that you can transform a fact into a fiction or a fiction into a fact just by adding or subtracting references. The effect of references on persuasion is not limited to that of 'prestige' or 'bluff'. Again, it is a question of *numbers*.[81]

Latour is claiming that the greater the number of references an article has, the less likely the conclusions made in the article will be challenged. Yet Latour's claim also implies that scientists actively engage in the practice of adding references and citations to their published works for the purpose of making them appear to be of a higher quality than they actually are, or to deceive the reader about the firmness of the conclusions drawn, or both. Latour's claim is also made by ignoring some well-grounded methods which operate in modern scientific research. Given the complexity of scientific disciplines today, it is not possible for any one researcher (or one group of researchers) to check individually each and every piece of information used. Therefore results from other researchers have to be taken provisionally as correct. Where are these results found? They are found in the relevant scientific literature and, if used, they must be referenced by some kind of citation.

Another reason for citing several references is for the purposes of corroboration. Experimental results must be repeatable if they are to carry any weight and the more times an experiment is repeated, the greater the likelihood is of finding error. The controversy over 'Cold Fusion' in 1989 is a prime example. Professors Fleischmann and Pons of the University of Utah, and Dr Jones and others, of Brigham Young University (also in Utah) claimed to have achieved nuclear fusion at room temperature. Nuclear fusion reactions in the Sun or in a hydrogen bomb require temperatures of millions of degrees. It was of no surprise to find that 'Cold Fusion' was greeted with much scepticism. This controversy was brought quickly to an end in the eyes of the majority of the physics community when nobody else was able to repeat successfully the Utah experiments. Where does a researcher find the results of repeated experiments? They too are found in the relevant scientific literature and, if used, must be referenced.

Latour goes to elaborate lengths in describing how scientists organize their articles and what tactics they employ in writing. We are told how they 'rally' the support of others, how they stack and posture their arguments, and how they encode hidden agendas that lead the reader in a direction which the author desires.[82] The purpose of all these manoeuvres is to halt those who oppose the acceptance of a given proposition as factual. (Latour's description sounds far more like a military exercise than the writing of a scientific paper!)

Latour tells us that a suitably structured scientific article with a multitude of references will be enough to stop all but the most determined of opponents. If such an opponent can get past the obstacles placed in his or her path by the literature and still challenge the validity of a fact, then he or she is brought face-to-face with even more formidable barriers found in the laboratory. Latour writes:

> When we doubt a scientific text we do not go from the world of literature to Nature as it is. Nature is not directly beneath the scientific article; it is there *indirectly* at best. Going from the paper to the laboratory is going from an array of rhetorical resources to a set of new resources devised in such a way as to provide the literature with its most powerful tool: the visual display. Moving from papers to labs is moving from literature to convoluted ways of getting this literature (or the most significant part of it).[83]

An opponent who manages to take his or her case of dispute all the way into the laboratory where the factual claim originated is presented with some kind of visual output from a scientific instrument. Visual displays are extremely powerful in their impact. In the originating laboratory an opponent is confronted with graphs or similar instrument outputs which cannot be questioned without also questioning both the workings of the instrument and the procedures involved in getting the instrument to produce its output.[84]

Latour is almost certainly correct about the compelling nature of visual displays. Consider an example from the Modern Revolution in geology. A magnetic seafloor profile is a visual representation of the direction of magnetization of strips of ocean floor. These profiles were used as evidence for the seafloor spreading version of continental drift. If the seafloor is created at oceanic ridges and spreads out on both sides of the ridge then there should be alternating directions of magnetization of the seafloor rocks as one moves out from the ridge on either side. The alternating directions correspond to past reversals of Earth's magnetic field. The resulting symmetric pattern of magnetization should be reflected in the magnetic profile. In 1966 a profile named Eltanin-19 proved to be utterly convincing evidence of seafloor spreading for most geologists. The impact of visual displays is illustrated by the following statement from one geologist upon viewing the Eltanin-19 profile: 'It's so good it can't possibly be true, but it is'.[85]

Latour says that the determined opponent who cannot find sufficient flaws in laboratory instrumentation and procedures is ultimately driven to build a counter-laboratory. In order to continue the dispute, the opponent must link up a network of black boxes and do so at an earlier stage in the relevant experimental process. This would allow the opponent an opportunity to question the original result since the instrumentation and experimental methods of the discoverer could be directly challenged.[86] Counter-laboratories offer the opponent a variety of new resources, not the least of which is the ability to produce the opponent's own visual displays. These displays can then be used to argue against the original claim.

In another of Latour's examples, he examines the debate between Pasteur and Pouchet in the 1860s over the issue of spontaneous generation of micro-organisms. This was a case of two opposing laboratories. Latour writes:

> The micro-organisms on which Pasteur depended were made to betray him: they appeared spontaneously thus supporting Pouchet's position . . . [Pasteur then] showed that the mercury used by Pouchet was contaminated. As a result Pouchet was cut off from his supply lines, betrayed by his spontaneous micro-organisms.[87]

Latour labels such a situation as one where the 'actors betray their representatives'. If we interpret the above passage literally, then it is just nonsense. If micro-organisms do exist then they are not the sort of entities that are capable of betraying anyone. If micro-organisms do not exist then there is nothing to do anything at all (including an act of betrayal). Neither Pouchet nor Pasteur was betrayed. A less literal interpretation of this passage of Latour's is that Pouchet's claim of spontaneous generation was challenged and overturned by Pasteur's laboratory. Pasteur was successful because he managed to secure sufficient resources and to deprive Pouchet of his own resources. The situation certainly can be depicted in this way but to do so is not to tell the whole story. Pasteur refuted Pouchet's claim by demonstrative and repeatable means. The outcome of Pasteur's experiment to show contamination of the mercury is not something which is, or can be, socially determined. After all, even Latour would have to admit that despite Pasteur's convictions and desires, his experiment could have failed.

This brings us to some very important questions which need addressing by constructivist sociologists. If scientific facts are socially constructed, then they have no real existence other than being in people's minds. Why then go to all the trouble of building huge laboratories? Why not save the expense and merely theorize about Nature in a manner similar to that of the ancient Greeks? Scientists and engineers build and operate laboratories with the professed aim of discovering new facts about the world. If scientific facts are merely social constructs, then either scientists are suffering from extreme cases of self-deception or they are totally dishonest about their research. In addition, we may raise the question asked in the previous chapter about the success of science. Why is science so successful if it is only socially constructed?

It is clear that Latour has identified some facets of the daily activities of scientists. However, most of these facets are not restricted to scientific research but are common to many fields of human endeavour (and all academic disciplines not just science). What Latour has done in presenting his argument is to exaggerate the extent of these aspects and to ignore or to play down other aspects of scientific research. Latour's description of science would be more appropriately applied to the workings of a criminal jury trial. In these trials, competing lawyers attempt to construct the guilt or innocence of an accused person in the minds of the members of the jury; each lawyer tries to get the numbers on his or her own side. The lawyers conduct their cases and play out their court-room tactics exactly as Latour describes. The conduct of scientific research is not equivalent to a legal battle or a military engagement but can be made to appear so if one is selective in the choice of aspects displayed.

In summarizing the impact of contemporary constructivist sociologists of science, Giere writes:

> While the critical reader may question whether it is "knowledge" or "the facts" that are being socially negotiated, it is undeniable that these works . . . capture the texture of day-to-day research in a way that few other works, be they sociological, historical, or philosophical, have ever done. Still, for anyone trained in the natural sciences or in an analytic philosophy of science, constructivism sounds wildly implausible.[88]

7
Rationality Revisited and Some Relations between General Theories and Individual Explanations

In this final chapter we shall address some questions that remain outstanding from earlier discussions. In particular, we shall attempt to shed some light on questions relating to the success of science and to scientific rationality: why is science so successful? Is it possible to account for science as an activity conducted in accordance with rational principles? Or is science only explicable under an analysis provided by the sociologist? The arguments in this chapter are presented in the hope that they will constitute a few more small pieces in the puzzle of science.

THE SUCCESS OF SCIENCE

Chapter 5 concluded with the question of why science is so successful if the motives of scientists are merely selfish. In Chapter 6 the question of the success of science arose again, this time in response to the claim that scientific facts are only social constructs.

Science not only works, it is highly successful at the empirical level. Yet if one accepts the claims of sociologists of science, then the fact that science is so successful is utterly unexplained. Why does science work? Scientific realism is the view that scientific theories are true (or nearly true); scientific theories give correct, or approximately correct, descriptions of real entities and processes that have an objective existence, independent of human observers. Those who adhere to a version of scientific realism (and there are many versions) are called 'Realists'. They claim that the success of science is due to scientific theories being true (or at least close to the truth). One prominent scientific realist is Hilary Putnam who has explicitly stated that any philosophy of science that does not embrace realism makes the success of science a miracle.[1] The realist point of view has great appeal.

Scientific realism cannot, however, be substantiated as a general thesis for the following reasons.

Just because a theory gives correct predictions this does not mean that the theory is true, or even approximately true. The empirical adequacy of a theory has no logical bearing on whether the theory is true or not. If one infers that a theory is true simply because it makes correct predictions, then one invokes a logical fallacy; one makes an error in reasoning. Logicians refer to this type of error as 'Affirming the Consequent'.[2] The error is made by assuming that if the conclusion is true then the premises upon which the conclusion is based must also be true. This does not necessarily follow since it is the case that a true conclusion can be derived from false premises: for example, the inferred conclusion might be true by sheer coincidence or for any number of other reasons.

Scientific theories do not need to be true, partially true, or even approximately true in order to be successful. Consider a standard example—the case of Newtonian physics. Newtonian physics is highly successful in predicting a wide range of phenomena. Yet if we accept our best confirmed theories, namely, relativity and quantum mechanics, as correct (or at least as being better approximations to reality) then we must concede that Newtonian physics is false. More specifically, we would say that the laws of Newtonian physics are false and that the ontology of Newtonian physics (absolute space, non-contact forces) is not real but a mere theoretical fiction. Despite all this, Newtonian physics is widely used, is taught in all senior school and tertiary courses in physics, and is the basis of many engineering practices and technologies. Why do we continue to use Newtonian physics if it is false? The answer is straightforward enough—it works! In the broad areas of its usage, calculations done with Newtonian physics give the same numerical results as would relativity or quantum mechanics if applied to the same problem and the Newtonian calculation is by far the easier to make. The theoretical structure of Newtonian physics is one example of a set of false premises that yield true conclusions.

Our second reason for not accepting scientific realism is that we have good inductive evidence from the history of science which indicates that most scientific theories are false. It is the case that all previously held scientific theories have (in some sense) been refuted or superseded. There is no guarantee that even our current theories will not be modified or replaced at some future time.

One might ask: are there not *some* true scientific theories? It is safe to say that there are some true theories in as much as we shall ever probably be able to establish their truth. Most of the true scientific theories we currently have are not terribly helpful though, because they tend to be either too general, too trivial or both. An example of such a true theory can be stated in six words: 'The planet Earth is roughly round'. This can be demonstrated by a variety of ground-based experiments and observations, but the results of such experiments did not stop people believing in a flat Earth for thousands of years. Today we have more direct evidence establishing the truth of this theory—to the extent which anything can be shown to be true. For example, one can now leave the surface of the Earth using some kind of rocketry transport and observe the whole Earth from a distance. However, just knowing that the Earth is roughly round does not, by itself, explain very much.[3] An analogy may be of assistance here—the story of the two clocks comes to mind. One clock is broken such that its hands point to the same numbers on the clock face at all times. The other clock runs such that it is always approximately correct but never reads the true time. The first clock gives the true time twice a day. However, nobody knows exactly when this occurs. The second clock never gives the true time. It only provides the approximate time and does so consistently. Which clock is more useful? Clearly it is the second clock, for what use is the first if we cannot tell when it is reading true. As with clocks, so too for theories. A theory need not be true in order to make correct predictions but to be particularly useful a theory must be empirically successful to a high degree of accuracy.

If scientific theories are not true, then how is it that they work and work so well? Several possible answers have been mooted. Ron Giere suggests that the success of science is due to causal interaction between scientists and the external world.[4] This is similar to an earlier idea of Bas van Fraassen, who writes:

> science is a biological phenomenon, an activity by one kind of organism which facilitates its interaction with the environment . . . I claim that the success of science is no miracle . . . For any scientific theory is born into a life of fierce competition . . . Only the successful theories survive—the ones which *in fact* latched on to actual regularities in nature.[5]

In this passage, van Fraassen is referring to empirical success. Whilst the level of empirical adequacy is of major importance to the assessment of any theory, it is not the sole consideration.[6] In cases of rival theories, which are underdetermined by the available data, empirical success does not assist in making a choice between these theories at all.

Larry Laudan has been a little more forthright in attempting to explain the success of science.[7] There is, of course, more to the success of science than mere empirical success. In order not to 'beg the question' of the success of science and to avoid his account being labelled as normative, Laudan outlines a set of goals for scientific research which are concerned with particular cognitive features. The success of science, he argues, can then be gauged on the basis of how well science can be said to have reached these cognitive goals. Laudan writes:

> These goal states concern themselves with certain interesting epistemic and pragmatic attributes. Consider a typical list of some of these aims:
> a) to acquire *predictive control* over those parts of one's experience of the world which seem especially chaotic and disordered;
> b) to acquire *manipulative control* over portion's of one's experience so as to be able to intervene in the usual order of events so as to modify that order in particular respects;
> c) to increase the *precision* of the parameters which feature as initial and boundary conditions in our explanations of natural phenomena;
> d) to integrate and *simplify* the various components of our picture of the world, reducing them where possible to a common set of explanatory principles.
> If we define cognitive 'success' along these lines, then it seems uncontroversial to say that portions of the history of science in the last 300 years have been a striking success story.[8]

Against such criteria as this, science has indeed been successful. The question is 'why?'. Laudan offers a more plausible explanation than Bas van Fraassen's proposal of theory elimination by 'natural selection'. Laudan's argument is that the various methods of theory testing and of theory selection employed by scientists are such that they produce reliable theories over time. Those theories which we come to call 'scientific' are efficient at advancing our cognitive aims and, in general, they do so better than theories we denote as 'non-scientific'.[9]

Laudan elaborates with some elementary examples. We shall consider one of these. Larry the philosopher has trouble starting his car one very cold morning. What should he do about this? The local garage is the answer, or so he believes. Larry thinks to himself that the mechanic at the garage will be able to fix the trouble by replacing the battery or perhaps the starter-motor. Larry leaves his car at the garage after explaining the trouble. On his return to the garage, Larry is amazed to find that the mechanic has replaced the car's brake drums! Larry is furious and promptly refuses to pay the mechanic. The mechanic is not pleased and points to the fact that the car now starts without any trouble. Larry points to the fact that since the work was done on the car the weather has warmed up considerably. The increase in atmospheric temperature and its effect on the car's engine constitutes Larry's explanation of why the car starts and has nothing to do with the (costly) replacement of brake drums. The mechanic says that he ascribes to a different theory. He believes that replacing the drums had the desired effect and this theory is supported by further evidence, namely, all the cars in his garage that had similar problems to Larry's also had their brake drums replaced and all now start.

In utter dismay, Larry responds by saying that, because he has a different explanation for the car starting, he insists that there be some tests designed which will help decide between their two theories and any others which are compatible with the available evidence. In addition, Larry says that he can find cars that had starting problems on the same day that his car did, but which now start without difficulty and did not have their brake drums replaced. This would, of course, add support to Larry's theory. Flippancy aside, Laudan concludes this example by stating:

> My mechanic's failure to impose any form of experimental controls on his causal claims is likely to lead him to make far less reliable predictions than I will. In short, my strategy will save me from several sorts of failure to which my mechanic friend will sometimes fall prey. This is not to say that hypotheses which pass my sorts of tests will never be mistaken, nor that theories which pass his tests will never lead to correct predictions. It is simply to say that my strategy will produce conjectures which break down less frequently and less quickly than his will, and this is precisely why we say that one theory is more successful than another.[10]

The methods of theory testing and sifting used in scientific research yield theories which although not necessarily true (or even close to being true) nevertheless assure a quality result, given enough time and effort, in accordance with our cognitive aims. These methods provide control procedures by which the theories generated are synchronized and calibrated to natural processes. 'Synchronization' means that the internal machinations of scientific theories are matched to, or are concurrent with, a range of natural phenomena. Yet this is not sufficient for empirical success. In order to be empirically successful, a theory must also be suitably 'calibrated', in the sense that the numerical predictions produced by a theory bear a close approximation to the measured values of physical quantities. In other words, the synchronization of a theory must be precise, or fine-tuned. It is important to note that there are two crucial factors in the design of successful scientific theories: the rigorous application of tried control and test methods and, most importantly, empirical (although theory-laden) input. Sociology of science, for the most part, does not take these factors into account. Instead the tendency of sociologists is to explain all scientific activity on the basis of social structures, social organization, social environment and various social interactions. The neglect of the epistemic and methodological dimensions of scientific research is a major failure of sociological theories of science.

SCIENCE IN A SOCIAL CONTEXT AND SCIENTIFIC RATIONALITY

Sociologists of science explain science as a social phenomenon that is not rational. In their view, science is an activity for which there are social goals, social standards and the outcome of scientific research is pieced together by social processes. It is quite obvious that scientific research is a social activity in so far as it is conducted by people and people are social creatures. Does the fact that people are social beings require that all human activities must be explained totally or even *predominantly* in terms of social psychology? Not necessarily! For instance, anyone who is hungry (and not otherwise constrained, for example, by religious beliefs) will seek out food. Such action is a rational and most natural response independent of social environment and cultural background. In more complex situations (including scientific contexts) why should there not also be rational explanations of behaviour?

What does it mean to say that an action is rational? In Chapter 2 (p. 52) a basic outline of rational action was presented: an action is rational if the individual performing the action has reasons for believing that the action will bring about, or assist in bringing to fruition, his or her aim (whatever that aim may be). To hold some reasons yet act against them is irrational, that is, such an action goes against reason. An action may not be fully rational but may not be irrational either if there are reasons, subjective ones, for doing it. The action is then a-rational.[11]

An account of rationality, or rational action, which is not concerned about the rationality of the agent's goals is referred to as 'Instrumental Rationality'. In such means–ends accounts of rationality, the aim is given and an action is rational if, on the available information, or even the agent's belief, this action optimizes reaching the goal. Newton-Smith describes instrumental rationality in the following terms:

> To generalize, to explain an action as an action is to show that it is rational. This involves showing that on the basis of the goals and beliefs of the person concerned the action was the means he believed to be most likely to achieve his goal.
>
> In this sense of rationality . . . the success of an explanation does not depend on the reasonableness of the goal. Neither does it depend on the truth and falsity of the beliefs in question, nor on their reasonableness or unreasonableness.[12]

Instrumental rationality does have explanatory utility—for

example, as used in Laudan's problem-solving model—but it does not give a total picture. If, for instance, the aims of the agent in question are themselves not rational then the activity of the agent can hardly be considered rational. One can think of any number of examples where an action appears rational relative to some goal (because the action optimizes the chances of attaining that goal) and yet the goal itself is not rational. Consider one such example. Suppose that my aim is to fly between London and Paris simply by vigorously flapping my arms. Relative to this aim, it would be rational to undertake many months of training and exercise in order to build up muscle, to reduce body fat and to improve blood circulation. However, no level of human fitness or any amount of personal organization is going to achieve this aim. Thus the question of the rationality of science must (at some stage) include the question of whether the aims of science are rational. This has been argued by several philosophers and denied by others.[13] Either way, a large part of the answer to the question of scientific rationality is to be found by employing instrumental rationality.

It is true to say that philosophers have been theorizing about the aims of science since at least the time of Aristotle. We noted some examples of cognitive aims of science in the previous section. Sir Karl Popper has also offered a few pertinent comments regarding the question of the aims of science:

> To speak of 'the aim' of scientific activity may perhaps sound a little naive; for clearly, different scientists have different aims, and science itself (whatever that may mean) has no aims. I admit all this. And yet it seems that when we speak of science we do feel, more or less clearly, that there is something characteristic of scientific activity; and since scientific activity looks pretty much like a rational activity, and since a rational activity must have some aim, the attempt to describe the aim of science may not be entirely futile.[14]

Different scientists do have different aims for their research. What is more, most aims are changed from time to time and for a variety of reasons. There are, however, some aims of science that are shared by most scientists and have consistently remained aims. We shall be concerned with two general aims that are not transient. The first is a short-term aim, that is, one that is realizable in the lifetime of a scientist. Short-term aims vary but there is at least one that is common to all researchers, namely, the solving of problems. This aim of science is emphasized in the theories of both Kuhn and Laudan. Whatever else individual scientists have as research objectives, they all attempt to solve some problem or other. Examples abound in science and range from fairly mundane objectives such as finding a more accurate value for Planck's constant, to finding a cure for Parkinson's disease. By virtue of conducting scientific research, all researchers have problem solving as an aim.

The second aim we shall be interested in is a long-term one. In the very long term (which may even be centuries) scientists aim at finding truths about the world. This is not a prescriptive stipulation, rather it is a descriptive claim. In the long term, scientists seek to discover truths or to contribute to such a process of discovery. The reader may have grave doubts about the validity of this, so let's be perfectly clear about what is and what is not being claimed here. It is not being claimed that a scientist's own immediate research efforts will (quickly) result in true theories, be they theories about quarks, or solar cells, or even genetic inheritance. The claim that science aims to find out truths is far, far more modest. Science aims for truth only in the sense that many individual scientists or groups of scientists, working over long time intervals contribute to a growing collective body of knowledge ('facts', theories) about the world and a small fraction of that body of knowledge is true. (Compare our earlier discussion (p. 173) that there are some true theories, in as much that we can know anything to be true.)

This, more modest, aim of finding truths is held by most scientists. When pressed on the matter, many scientists will admit to accepting this as a long-term aim for science as a whole. The obvious objection to this is that, just because scientists may agree that the finding of truths is an aim of science, their admissions alone do not make it so. Scientists might, for instance, publicly advocate one or more aims of science whilst privately, or even unconsciously, adhering to others. However, there are some good reasons for accepting the finding of truths as an aim for science. Consider three such reasons: more than any other expressed aim of science, the search for truths captures the essence of human curiosity about the natural world and the desire to understand things as they are; we have corroborating evidence to the effect that science has indeed discovered truths about the world (even the hardest-nosed anti-realist would surely not deny that the Earth is roughly round!); since the finding of truths is a very long-term aim, the knowledge arrived at is not bound by immediate limitations placed upon institutionalized scientific research, particular research groups, or individual scientists. Neither need the resulting research be influenced by social, political or religious pressures. Acceptance of the finding of truths as a long-term aim of science does not entail full-blown scientific realism, but neither does it imply that this aim is entirely unproblematic.

There is a definite relationship between these two general aims of science. Aside from its pragmatic benefits, the short-term aim of problem solving is seen as a necessary condition for the achievement of the long-term aim of gaining truths about the world. Why is this the case? We have found no other way of investigating our world that yields the same sort of results as does scientific research. Moreover, it is to be fully expected that a true theory, along with suitable boundary conditions, will be the best problem-solver of all possible alternatives. We know that the physical world and the human condition together place restrictions on our epistemic access to the world. Granting that we do have epistemological restraints, the short-term aim of problem solving is a necessary stepping-stone to achieving the long-term aim of finding truths. It is because of this relationship between these two aims that problem solving itself gains legitimacy as a rational aim. Therefore we need only restrict our discussion to instrumental rationality where the aim is problem solving.

EXTENSIONS TO LAUDAN'S THEORY

Laudan's theory has demonstrated an ability to make rational sense of many more historical instances of theory choice and theory change than earlier theories of science. In *Progress and its*

Problems, Laudan summarized his position on the rationality of theory choice:

> how can we . . . continue to talk normatively about the rationality (and irrationality) of theory choices in the past, while at the same time avoid the grafting of anachronistic criteria of rationality onto those episodes?
>
> The [problem-solving] model I have outlined resolves part of that difficulty by exploiting the insights of our own time about the *general* nature of rationality, while making allowances for the fact that many of the *specific* parameters which constitute rationality are time- and culture-dependent. It transcends the particularities of the past by insisting that for all times and for all cultures, provided those cultures have a tradition of critical discussion (without which no culture can lay claim to rationality), rationality consists in accepting those research traditions which are most effective problem solvers . . . the model argues that there are certain very general characteristics of a theory of rationality which are *trans-temporal* and *trans-cultural* . . . the model also insists that what is specifically rational in the past is partly a function of time and place and context.[15]

We now introduce the term: 'meta-level methodology' (or just 'meta-methodology' for short). A meta-methodology advises which methodology is appropriate to employ; it is a rule, or set of rules, that determine the choice of one's methodology. The problem with proposing the existence of any meta-methodology is that it can result in an infinite regression of higher-level methodologies. If a meta-methodology is needed to choose between methodologies, do we then need a meta-meta-methodology for choosing our meta-methodology? Do we then need a meta-meta-meta-methodology for choosing our meta-meta-methodology and so on *ad infinitum*? What would be better is a scheme which does not extend beyond the first meta-level.

In a later book, *Science and Values*, Laudan outlines what he calls the 'Reticulated Model of Scientific Rationality', which essentially plays the role of a weak meta-methodology. This model is designed to display the inter-dependence between aims, methodologies, factual claims and their roles in bringing about consensus in science. All three are important components of every research tradition and all three may be subject to change during the life of a research tradition. In the reticulated model, these three components are also viewed as levels of scientific commitment. The 'axiological' level concerns cognitive aims of science. The 'factual' level is concerned with both facts and theories. The model displays a 'triadic network of justification' in which each component is mutually dependent in the following ways. Methods are justified by appeal to shared (or common) aims. Theories are justified by appeal to accepted methods. Theories act as a constraint upon which methods are acceptable. Methods indicate whether one's aims can be realized or not, and one's theories must harmonize with one's aims. Laudan explains:

> The reticulational approach shows that we can use our knowledge of available methods of inquiry as a tool for assessing the viability of proposed cognitive aims . . . the reticulated picture insists that our judgements about which theories are sound can be played off against our explicit axiologies in order to reveal tensions between our implicit and our explicit value structures . . . there is a complex process of mutual adjustment and mutual justification going on among all three levels of scientific commitment . . . Axiology, methodology, and factual claims are inevitably intertwined in relations of mutual dependency.[16]

We will only deal with those aspects of the reticulated model that are relevant to our present considerations of meta-methodology, except to note briefly some comments on the model. Le Grand finds evidence from the history of science in favour of it. Worrall thinks that it is wrong-headed. Brown believes that, although the model is on the right track, the situation in actuality is more complex than Laudan depicts.[17] What is particularly relevant in the reticulated model to our discussion is the mechanism whereby one's aims justify one's methodology which is, in turn, constrained by one's theories. This point will be returned to shortly.

In his 1969 Postscript to *The Structure of Scientific Revolutions*, T. S. Kuhn writes:

> A scientific community consists . . . of the practitioners of a scientific specialty. To an extent unparalleled in most other fields, they have

> undergone similar educations and professional initiations . . . Communities in this sense exist, of course, at numerous levels. [Small scientific] communities [are] of perhaps one hundred members, occasionally significantly fewer . . . Communities of this sort are the units that this book has presented as producers and validators of scientific knowledge. Paradigms are something shared by members of such groups.[18]

Kuhn's purpose in re-defining a scientific community (or perhaps micro-community is a better word) was to avoid some of the criticisms levelled at his original theory. In particular, there need no longer be complete consensus within a broad scientific discipline; different groupings (or micro-communities) may disagree over issues.[19] Kuhn also thinks that there may be small scientific revolutions (micro-revolutions) in the time interval between large-scale scientific revolutions. Such small revolutions would not necessarily require a crisis state to bring them about.[20] If Kuhn was writing *Structure* from scratch today, one might get the distinct impression that specific scientific interests and the narrow professional education of small specialist groups of scientists would be major, and sometimes determining, factors in both theory choice and the dynamics of theory change.

In an attempt to improve problem-solving models of science, Le Grand has suggested that these models should explicitly take into account such scientific interests.[21] Although Laudan's original theory offers the promise of rationally accounting for theory choices of most scientists, it stops there. Laudan's theory is in need of some form of supplementation or modification. Le Grand finds Laudan's model preferable to its alternatives but finds it necessary to employ interest analysis in order to explain why a minority of geologists and geophysicists did not follow the majority decision against continental drift in the 1920s. The explanation for the action of this minority stems from their specialist interests and a greater appreciation of palaeomagnetic and palaeoclimatological data. Le Grand's position is summarized as follows:

> for most geologists, it [continental drift] offered no advantages over its competitors and was not pursued . . . localism can play important

> roles in the reception of novel theories and the appraisal of rival programs. Further, if one theory be accepted even by a minority of the practitioners in one specialty and another in other specialities, the effect is to spread the risk and perhaps to maximize progress through competition.
>
> This amounts to an extension or refinement of the 'scientific' or 'technical interests' invoked by some social analysts of science and its grafting onto problem-solving models.[22]

('Localism' refers to a detailed examination of a small, specific area of research rather than 'global' theories.)

In a similar vein to Le Grand's supplement, Andrew Lugg has proposed a modification to Laudan's problem-solving model, which is primarily targeted at explaining disagreement in science as a rational process. Lugg argues that scientists have 'access differences', that is, differences in access to the total system of scientific beliefs and practices.[23] These differences, says Lugg, account for disagreements amongst scientists and that such disagreement is not only to be expected, it is entirely rational. The existence of access differences also assists in accounting for different theory choices made by small groups of researchers. Lugg writes:

> if science is indeed appropriately pictured as comprising an enormous network of scientific beliefs and practices to which scientists have varying and incomplete access, a natural way of modifying Laudan's account suggests itself: we should relativize the choice of a theory to the chooser's location with respect to this system. When we do this we obtain a model of scientific rationality which recognizes the possibility of scientists' evaluating the problem-solving effectiveness of competing theories in different ways and hence the possibility of rational scientists disagreeing about their relative acceptability. If scientists have different access to the total system of scientific beliefs and practices, we can expect them to differ with respect to what counts as an empirical problem, the sorts of objections they recognize as conceptual problems, the criteria of intelligibility, the standards of experimental control, the importance of weight they assign to problems, and so on.[24]

The inclusion of both 'Specialist Interests' and 'Differential Access' in rational models of science provides reasons for the decisions of individual scientists or small groups of researchers. The explanatory advantage gained by these extensions is that they account for situations where a small, specialist group of scientists assigns weightings to various pieces of data, which are different from those which other groups assign to the same data. A certain piece of data may be judged by one specialist group as strong evidence for a particular theory, whilst another group may not regard this data in anything like the same light. A consequence of specialization and differential access is that there is a tendency, in those researchers concerned, to have their theory choice pre-structured (compare the discussion in Chapter 6, p. 154). Another way of stating this point is to say that specialization and differential access work to 'inform' one's methodology. This then leads us back to meta-methodological considerations. The aim of furthering one's specialist interests (in part) justifies one's methodology and what's more, this does not entail an infinite regression of meta-levels. Differential access will also tend to steer individual researchers in directions of theory choice that appear more illuminating by virtue of the researcher's particular access to the network of scientific beliefs.

We noted in the previous chapter (pp. 153–4) that scientific interests may be due to, or reinforced by, larger social interests. This being the case, it can be seen that there is a distinctive role for social factors in helping to shape one's meta-methodology and a role for the sociologist of science in identifying these factors. This does not, however, lead to the conclusion that theory choice is irrational or entirely socially determined. The present analysis indicates that the models of rationality required to depict accurately the scientific enterprise are far richer than was traditionally imagined by philosophers.

GENERAL THEORIES VERSUS INDIVIDUAL EXPLANATIONS AND THEIR CONNECTIVE LINKS

This section concerns the relations between a general explanation —an explanation that accounts for a majority decision—and explanations of individual actions or choices. A general explanation cannot be assumed to be identical to, or merely the sum of, individual explanations. In most circumstances where a group of people perform an action or make a choice there are usually a number of explanations for performing that action or making that choice. Some individuals in a group may do or act as others in the group but for their own particular reasons and these reasons differ from those held by the majority. There may also be those in the group who decide not to act as the majority of group members do. If the individuals who dissent from the group decision do so for their own strong reasons, then their decisions may be considered to be rational ones. Such individuals we shall call rational dissenters. If, on the other hand, there are members who dissent from the group decision but for the same reasons as the group has used in coming to their decision, then the decision of the dissenters may well be considered an irrational one. We shall not be concerned here with this sort of irrational choice. This leaves the dissenters who have alternative reasons and those who accept the group decision but for their own particular reasons.

We shall concentrate on the relations, or connective links, between individual reasons and the general (group) explanation where the action or choice is the same. In such cases the proposal is that the following connective links exist between general and individual explanations:

(I) both types of explanation (individual and general), although different, lead to a common result;

(II) the general explanation will be identical to either all or a majority of the individual explanations;

(III) where general and individual explanations are not identical, the general explanation may still account for some of the individual ones.

What is to be done if the independent explanation is inconsistent with the general explanation? In such cases, the following 'Consistency Condition' shall be imposed:

> When the general and an individual explanation are not identical and the general explanation does not account for the individual one, then this individual explanation does not act as a refuting instance of the general one.

How can this consistency condition be valid if the individual explanation is really inconsistent with the general case?

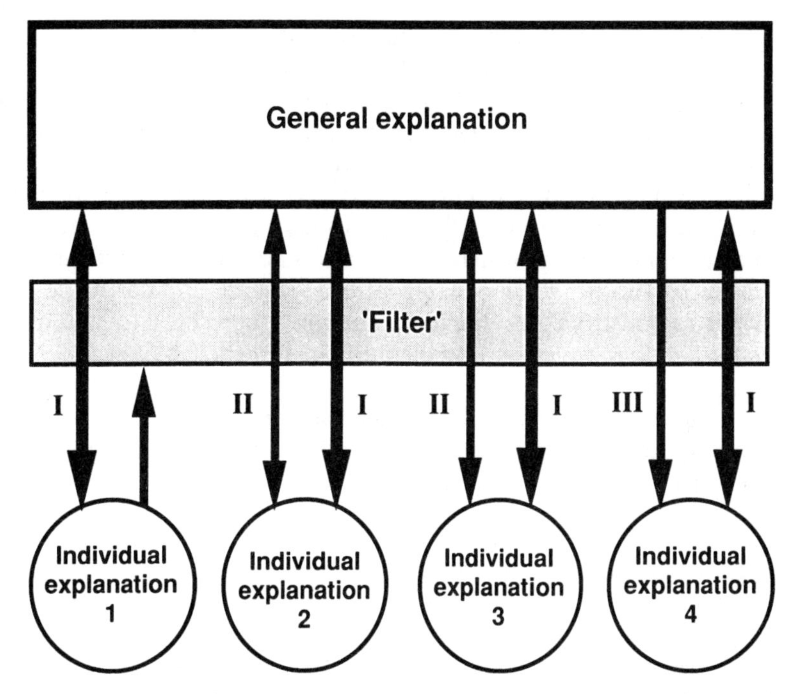

Figure 6 The relations between a general theory (or explanation) and individual explanations

The consistency condition is acceptable if the alleged refuting instance is somehow screened-off from the general explanation. Consider Figure 6. In this diagram the links between the general explanation and four individual explanations of the same action and/or choice are represented by arrows labelled I, II or III, depending on their type. Individual explanations 1 and 4 are not identical with the general explanation. Explanations 2 and 3 are identical with the general explanation. The thick, double-headed arrows marked I are shared between all individual explanations and the general explanation, as this indicates that all are linked through leading to a common result. The double-headed arrows marked II represent the identity link, that is, where the general and individual explanations are the same. The single-headed arrow marked III indicates a link where the general explanation accounts for the individual one but not vice versa (thus the single arrowhead in its representation).

Between the general and the individual explanations is what shall be called a 'filter'. This filter represents a mechanism by virtue of which the consistency condition holds. The exact mechanism will be case dependent. The primary function of this filter is to ensure that individual explanations that are prima facie inconsistent with the general explanation do not act as refuting instances of the general explanation. The unmarked single-headed arrow from individual explanation 1 represents a directed *modus tollens* which is screened-off or absorbed by the filter mechanism.

A fairly simple example serves to illustrate these points. A typical piece of market research might be conducted to find the answer to a question such as:

What is the basic reason people buy product X?

Implicit in this question are the assumptions that people buy product X and that there is a basic reason why people buy the product, rather than their buying it for no particular reason. We shall take these initial assumptions for granted. Suppose that a representative sample of consumers is surveyed and all reply with the same answer:

I buy product X because I need it!

This also turns out to be the only reason why they buy it. The market researcher's conclusion is:

People need product X.

Let's further suppose that after the survey was completed the researcher who conducted it decides (out of curiosity) to interview his next-door neighbour named Sally. Now Sally gives a totally different answer. She says:

I buy X because I *like* it!

Sally is then asked if she also needs product X, since needing the product would be a more basic reason for buying than merely liking it. Sally answers in the negative; she can do without product

X but prefers not to. Here we have a simple example where an individual explanation differs from what was thought to be *the* explanation, namely:

All people who buy product X buy it because they need it.

The particular instance of Sally's behaviour indicates that this explanation is not correct. Clearly, not all people who buy product X buy it because they need it. The presence of the universal quantifier 'all' in this statement renders it false. Therefore a suitable alteration might be to replace 'all' with another quantifier, say 'most'. If this is done then we have:

Most people who buy X buy it because they need it.

This is now a true statement given that the sample interviewed was representative of the buying population.

This latter explanation is not, however, an answer to the original question posed, that is, to this question:

What is the basic reason people buy product X?

The original question makes no reference to terms such as 'most' or 'majority'. It might be claimed that this question assumes that there is only one basic reason people buy the product. It might also be asserted that there is simply no answer to the above question. What is plainly obvious is that there is more than one reason for buying product X, for Sally buys for one reason (she likes it) and everyone else buys it for one other reason (they need it). What then qualifies as an answer to the question of *the* basic reason for buying product X, if indeed there is an answer to this question?

The basic reason, and indeed the one and only reason, all people (except for Sally) buy product X is that they need it. We should also say that, since Sally's reason for buying X is that she likes it, this reason is basic for her because it is her only reason. But not only is Sally's reason not basic for everybody else, it is not even a reason for everyone else. Consequently, Sally's reason is *inapplicable* to everybody else. In addition, needing is more basic than liking, as is evident by appeal to specific examples such as physiological ones. (I may like chocolate but I need food. If I can have chocolate as part of my food intake, this is all very nice, but my basic need can be fulfilled without chocolate.) Even though Sally does not need product X, her liking it cannot be considered as a basic reason for purchasing it because her reason is not basic when weighed against everyone else's reason for buying X. It can now be asserted that the answer to the original question is:

The basic reason people buy product X is that they need it.

No inconsistency is involved in asserting this response. What this example purports to illustrate is that if an individual explanation appears, prima facie, a counter-instance to a general explanation, it does not necessarily refute that general explanation.

What can be identified as the filter mechanisms in this example? (Recall that the function of the filter is to absorb all directed *modus tollens*.) Firstly, we saw that Sally's reason for buying product X is not a reason for anyone else buying X. Thus on the grounds of *non-applicability*, her reason for buying is ruled out. Secondly and more importantly, her reason is not a basic one for buying X when the whole set of buyers is considered. Sally's reason is therefore *contextually* excluded. These are two filtering mechanisms and, although related, they are still logically distinct.

CASE-STUDY ON THE APPLICATION OF 'FILTER' MECHANISMS

Let's apply the ideas outlined in the previous section to a case of scientific theory choice. We take up again the historical case in which the Anglo-American geological community had to choose between two rival theories: permanentism and Wegener's version of continental drift. Why choose this for a case-study? It is a relatively recent case in the history of science and has received attention from many historians of science, philosophers of science, philosophically oriented historians, and sociologists.[25] Thus there is a mass of material already available on this case and a number of different interpretations of that material. The outcome of the debate in the 1920s over the two theories is well known. Drift was not chosen. Despite this, continental drift theory was triumphant in the 1970s. The major forum for the debate was the 1926 New York Symposium on Continental Drift. Takeuchi, Uyeda and Kanamori claim the following reasons led to the rejection of Wegener's theory:

- no adequate mechanism was proposed to move the continents;
- the apparently more solid rocks of the Earth's mantle must give way to those of the crust;
- the timing of the initial continental break-up, as given by Wegener, was considered to be far too recent in the Earth's history;
- the trans-Atlantic fit was not as good as Wegener claimed;
- the proportions of geological features and species similarities across the Atlantic were not as high as would be expected if Drift were a reality.[26]

In Laudan's problem-solving model, a rational choice is made between competing research traditions by a criterion of problem-solving adequacy, determined in turn by a tradition's problem-solving effectiveness. The rational choice between rival traditions is made by choosing the most effective problem-solver. It has been argued (for example by Frankel)[27] that in terms of Laudan's criterion of adequacy, Wegener's version of drift was not a more effective problem-solver than was permanentism. Most geologists could be said to have decided against drift for the reasons given above. These reasons readily lend themselves to evaluation by Laudan's criterion and we can safely hold that (for those that did so) the rejection of continental drift for these reasons was a rational choice. There were, however, several other reasons for rejection either made vocal or implied at the 1926 symposium. We shall consider three of these alternative reasons which are attributable to leading American academic geologists of the time: Edward Berry, Charles Schuchert and R. T. Chamberlin.

Berry rejected continental drift primarily on methodological grounds:

> My principal objection to the Wegener hypothesis rests on the author's method. This, in my opinion, is unscientific, but takes the familiar course of an initial idea . . . and ending in a state of auto-intoxication in which the subjective idea comes to be considered as an objective fact.[28]

This explanation of Berry's choice can be identified as being of the type of individual explanation where the general explanation (or theory) accounts for the individual. Using a Laudanian analysis, Berry's objection would pose an external conceptual problem for Wegener's version of drift. Recall that an external conceptual problem is where a theory conflicts either with an already accepted theory, or with some other strongly held belief.[29] Even if this reason provided the only explanation for Berry's rejection of drift, Laudan's theory can still be used to account for it. Wegener's methodology constituted a major conceptual problem for the continental drift research tradition. On the basis of the (heavy) weighting that Berry assigned to this conceptual problem, it was a rational decision for him to reject drift.

Schuchert's main reason for rejecting Drift is summed up by Le Grand:

> he [Schuchert] argued that the continents do not fit unless they are distorted from their modern shape. This is evidence against Wegener. But, if they did fit, this too would be evidence against Wegener because no one could believe that the continents ploughed undistorted through the ocean floor . . . Wegener is damned if they fit and damned if they don't![30]

Surely we cannot entertain Schuchert's explanation for rejecting drift as rational, for the simple reason that his argument is inconsistent. He cannot have it both ways! Schuchert's explanation must be excluded on the grounds of its lack of consistency. This is the filter mechanism in the case of Schuchert's individual explanation.

Lastly, we consider what is perhaps the most interesting of the three cases—R. T. Chamberlin's explanation. His main reason for rejection was more implied at the 1926 symposium than explicit, but had been voiced some years earlier. Chamberlin commented that if Wegener's hypothesis were taken seriously then geological studies of the previous seventy years would have to be abandoned and a fresh start made.[31] This consequence of accepting drift was well known to those attending the 1926 symposium, even if it was not publicly stated at that time. Le Grand explains why this can be inferred:

> The passion and direction of the rhetoric hurled at Wegener suggests that some of his opponents saw more at stake than the question of which global theory was best. Clues to these issues are provided . . . by reading between the lines. If Drift be accepted, much of the geological theory would become outmoded.[32]

If we take this as the primary explanation of Chamberlin's rejection of drift, then sociologists could claim it to be a counter-instance to Laudan's general explanation of theory choice. It does seem that we have here a clear case of a choice made for reasons other than those given by the general explanation and which cannot be accounted for by the general explanation.

How are we to deal with this? A first response might be to say that if Chamberlin's reason for rejecting drift is either his alone or is a primary reason for only a small minority of the Anglo-American geological community, then this situation is, in part, a parallel to the market research example (pp. 188–90). In other words, if Chamberlin's reason is not a reason for theory rejection (as distinct from being a reason for professional uneasiness) for the majority of geologists then we can exclude it on the grounds of non-applicability, in a similar fashion to the procedure used in the market research example. This then is one possible filtering mechanism for absorption of the directed *modus tollens*. If Chamberlin's reason did not constitute a primary reason for the rejection of drift by a majority of the community, we might well exclude it on contextual grounds, again in a manner similar to the market research example.

Rational dissenters were defined earlier (p. 186) as those who decided not to accept the group decision and did so because they had strong reasons of their own for such dissension. An account was given of the actions of rational dissenters by appeal to specialist interests and differential access. The filter argument outlined above indicates how a rational account of theory choice need not be subject to refutation by merely pointing to apparent counter-examples. There are other implications from this analysis. When the use of a filter mechanism is appropriate, this indicates that a sociological explanation of the action is required because the action is not fully rational (such as in Schuchert's case). The need for a sociological explanation when 'all else fails' was the crux of Laudan's 'A-rationality Assumption'.[33] This assumption has been criticized mainly by sociologists (who dislike being told what should and should not be their domain of study).[34] However, we have already noted that the domain of the sociologist of science should not be restricted to the 'sociology of error' (to use Bloor's phrase) but should also include the role of identifying those social factors which assist in shaping meta-methodologies.

There is another implication arising from the filter argument. We saw that the exact filter mechanism used was case dependent. In other words, we used different explanations to account for rational and irrational choices. Since this usage runs counter to the symmetry tenet of the strong programme in the sociology of scientific knowledge, additional doubt is cast on the validity of that tenet.[35]

PROSPECTS FOR THE RATIONALITY OF SCIENCE

The rationality of science, contrary to the claims of some sociologists, is alive and well. A full investigation and extension of existing rational models of science has, by no means, been exhausted by philosophers. There is a bright future for richer accounts of scientific rationality, accounts which take full advantage of both the cognitive and social aspects of scientific research. There is a need for the rationality of science to hold its own ground to balance out the explosion of social accounts (as Woolgar has described it). Avenues for continuing research are already being pursued in the forms of naturalized epistemology and in cognitive science studies.

Notes

1 SCIENCE AND ITS PHILOSOPHY

[1]Feyerabend, 'Philosophy of Science: A Subject with a Great Past'.
[2]Passmore, 'Logical Positivism', pp. 53–5.
[3]Popper, *The Logic of Scientific Discovery*, section 9.
[4]For a more formal introduction to inductive arguments, see Skyrms, *Choice and Chance*.
[5]Russell, *Problems of Philosphy*, pp. 63–4 (his italics).
[6]Losee, *A Historical Introduction to the Philosophy of Science*, p. 61.
[7]Aristotle, *Prior Analytics*, II, 23, 68b.
[8]Bacon, *Novum Organum*, I, 105.
[9]These are not to be confused with the (different) Forms postulated by Aristotle's teacher Plato (428–348 B.C.).
[10]Bacon, *Novum Organum*, I, 19.
[11]Ibid., I, 39.
[12]Ibid., I, 41–51, and Losee, *A Historical Introduction*, p. 62.
[13]Bacon, *Novum Organum*, I, 53.
[14]Ibid., I, 61, and Losee, *A Historical Introduction*, p. 62.
[15]Cranston, 'Francis Bacon', pp. 238–9.
[16]Losee, *A Historical Introduction*, pp. 64–5, and *Novum Organum*, II, 11–19.
[17]Rossi, *Francis Bacon*, pp. 219–20.
[18]Ibid., p. 222.
[19]For a summary of some attempts to solve the problem of induction, see Salmon, 'The Foundations of Scientific Inference'.
[20]Blake, Ducasse and Madden, *Theories of Scientific Method*, p. 71.
[21]Rossi, *Francis Bacon*, pp. 220, 221.
[22]The term 'inductivism' is also used in another sense. It can denote a position which emphasizes the significance of induction to scientific development. Cf. Losee, *A Historical Introduction*, Chap. 10.
[23]Scheffler, *Science and Subjectivity*, p. 7.
[24]Ibid., p. 8.
[25]Ibid., p. 9.
[26]See also Bunge, *The Myth of Simplicity*.
[27]Quine, 'On Simple Theories of a Complex World', p. 255.
[28]The simplicity requirement in the specification of induction is discussed at length by Clendinnen in 'Rational Expectation and Simplicity'.
[29]Gregory, *Eye and Brain*, pp. 204–15.
[30]Turnbull, *The Forest People*, pp. 227–8.
[31]Leicester, *The Historical Background of Chemistry*, p. 137.
[32]For further examples, see Hanson, *Perception and Discovery*, esp. Chaps 10–11.
[33]Smart, *Between Science and Philosophy*, p. 80 (his italics).
[34]See also Hesse, 'Is There an Independent Observation Language?'.

2 KUHN'S THEORY OF SCIENTIFIC REVOLUTIONS

[1]Kuhn, *The Structure of Scientific Revolutions*, p. vii.
[2]Ibid., p. 4.
[3]Ibid., pp. 12–13.
[4]Oldroyd, *Darwinian Impacts*, p. 200.
[5]Kuhn, *The Structure of Scientific Revolutions*, p. 17.
[6]Ibid.
[7]See Home, 'Franklin's Electrical Atmospheres'.
[8]Kuhn, *The Structure of Scientific Revolutions*, pp. 18–19.
[9]Ibid., p. 10.
[10]Ibid.
[11]Ibid.
[12]Ibid., p. 23.
[13]Ibid., p. 24.
[14]Ibid., p. 37.
[15]Ibid., pp. 25–7, 34.
[16]Ibid., p. 38.
[17]Ibid., p. 42.
[18]Ibid., p. 47.
[19]Ibid., p. 52.
[20]'The Nature of Paradigm'.
[21]Kuhn, *The Structure of Scientific Revolutions*, p. 81.
[22]Ibid.
[23]Ball, *A Short Account of the History of Mathematics*, p. 374.
[24]Kuhn, *The Structure of Scientific Revolutions*, p. 55.
[25]Ibid., p. 150.
[26]Ibid., p. 52.
[27]Ibid., pp. 52–3.
[28]Ibid., p. 59.
[29]See Popper, *The Logic of Scientific Discovery*, Chap. 2.
[30]Kuhn, *The Structure of Scientific Revolutions*, p. 146.
[31]Popper, *Conjectures and Refutations*, pp. 242ff. See also pp. 63–5 for further details.
[32]Kuhn, *The Structure of Scientific Revolutions*, p. 81.
[33]Sciama, *Modern Cosmology*, pp. 7–8.
[34]Wolfenstein and Beier, 'Neutrino Oscillations and Solar Neutrinos', pp. 28–9.
[35]Kuhn, *The Structure of Scientific Revolutions*, p. 82.
[36]Pias, *'Subtle is the Lord ...'*, p. 140.
[37]Dreyer, *A History of Astronomy from Thales to Kepler*, pp. 290, 309.
[38]Kuhn, *The Structure of Scientific Revolutions*, p. 82.
[39]Ibid., p. 77.
[40]Ibid., p. 79.
[41]Leicester, *The Historical Background of Chemistry*, p. 125.
[42]Kuhn, *The Structure of Scientific Revolutions*, pp. 90–1.
[43]Ibid., p. 84.
[44]Ibid., p. 68.
[45]Ibid., p. 79.
[46]Ibid., p. 5.
[47]Ibid., pp. 64–5.

[48]For further examples, see Hanson, *Perception and Discovery*, Chaps 5–6 and Gregory, *Eye and Brain.*
[49]Kuhn, *The Structure of Scientific Revolutions*, pp. 89–90.
[50]Ibid., p. 90.
[51]Pias, *'Subtle is the Lord . . . '*, p. 47, Chap. 7.
[52]Kuhn, *The Structure of Scientific Revolutions*, p. 122.
[53]Ibid., pp. 151–2.
[54]Drake, *Galileo*, p. 44.
[55]Kuhn, *The Structure of Scientific Revolutions*, p. 94.
[56]Ibid., p. 112.
[57]Ibid., pp. 148–50.
[58]There are many articles on the incommensurability of theories and the non-translatability of theoretical terms. For example, see Biagioli, 'The Anthropology of Incommensurability', and Sankey, 'Translation Failure Between Theories'.
[59]Kuhn, *The Structure of Scientific Revolutions*, p. 85.
[60]Ibid., p. 79.
[61]Ibid., p. 152.
[62]Leicester, *The Historical Background of Chemistry*, p. 137.
[63]Kuhn, *The Structure of Scientific Revolutions*, p. 94.
[64]Ibid., pp. 152–3.
[65]Takeuchi, Uyeda and Kanamori, *Debate About the Earth*, p. 72.
[66]Kuhn, *The Structure of Scientific Revolutions*, p. 103.
[67]Ibid., p. 137.
[68]Ibid.
[69]Orwell, *Nineteen Eighty-Four*, p. 186.
[70]Kuhn, *The Structure of Scientific Revolutions*, p. 137.
[71]Musgrave, 'Why did oxygen supplant phlogiston?', p. 191.
[72]Ibid., pp. 195–6.
[73]Kuhn, *The Structure of Scientific Revolutions*, pp. 170–1.
[74]For a fuller discussion of these issues, see Barnes, *T. S. Kuhn and Social Science.*
[75]Ezrahi, 'The Political Resources of Science', p. 222.
[76]Hunt and Sherman, *Economics*, p. 117.
[77]Ibid., p. 119.
[78]Iltis, 'Leibniz and the Vis-Viva Controversy', p. 21.
[79]Iltis, 'The Leibnizian-Newtonian Debates', p. 358.
[80]Ibid., p. 365.
[81]Laudan, 'The *Vis viva* Controversy, A Post-Mortem', p. 131.
[82]Iltis, 'Leibniz and the Vis-Viva Controversy', p. 32.
[83]Iltis, 'The Leibnizian-Newtonian Debates', p. 365.
[84]Feyerabend, 'Consolations for the Specialist', p. 22.

3 LAKATOS'S METHODOLOGY OF SCIENTIFIC RESEARCH PROGRAMMES

[1]Lakatos, 'Criticism and the Methodology of Scientific Research Programmes' and 'Falsification and the Methodology of Scientific Research Programmes'. (All references to this latter article will cite page numbers from Lakatos and Musgrave (eds), *Criticism and the Growth of Knowledge.*)

[2]Kant, *The Critique of Pure Reason* (1855).
[3]Lakatos, 'Falsification and the Methodology of Scientific Research Programmes', pp. 94–5.
[4]Ibid., p. 174.
[5]Lakatos, 'Introduction: Science and Pseudoscience', pp. 6–7.
[6]Lakatos, 'Falsification and the Methodology of Scientific Research Programmes', p. 99.
[7]Ibid., pp. 99–100 (his italics).
[8]Popper, *The Logic of Scientific Discovery*, p. 111.
[9]Lakatos, 'Falsification and the Methodology of Scientific Research Programmes', p. 106.
[10]Ibid.
[11]Ibid., p. 107.
[12]Ibid., p. 92.
[13]Ibid., p. 117.
[14]Ibid., pp. 138–40.
[15]Ibid.
[16]Ibid., p. 116.
[17]See, for example, Pannekoek, *A History of Astronomy*, p. 358.
[18]Lakatos, 'Falsification and the Methodology of Scientific Research Programmes', p. 116, n. 4.
[19]Ibid., p. 117.
[20]Pannekoek, *A History of Astronomy*, p. 141ff.
[21]Lakatos, 'Falsification and the Methodology of Scientific Research Programmes', pp. 117–18 (his italics).
[22]Grosser, *The Discovery of Neptune*, pp. 100–1.
[23]Lakatos, 'Falsification and the Methodology of Scientific Research Programmes', p. 118 (his italics).
[24]Ibid., pp. 118–19.
[25]Ibid., p. 118.
[26]Ibid., p. 119, n. 2.
[27]Ibid., p. 133.
[28]Wegener, *The Origin of Continents and Oceans*; Takeuchi, Uyeda and Kanamori, *Debate About the Earth*, Chaps 8, 9.
[29]Lakatos, 'Falsification and the Methodology of Scientific Research Programmes', p. 130.
[30]For a more formal introduction to the rule of *modus tollens* and some examples, see Lemmon, *Beginning Logic*, p. 12.
[31]Lakatos, 'Falsification and the Methodology of Scientific Research Programmes', p. 133.
[32]Ibid., p. 135.
[33]Ibid., p. 136 (his italics).
[34]Ibid., pp. 136–7 (his italics).
[35]Ibid., p. 120.
[36]Ibid., p. 155.
[37]Frankel, 'The Career of Continental Drift Theory', pp. 58–9.
[38]Lakatos, 'Falsification and the Methodology of Scientific Research Programmes', p. 133.
[39]Ibid., p. 134 (his italics).
[40]Ibid., p. 137.
[41]Musgrave, 'Method or Madness?', p. 457.

[42]Lakatos, 'Falsification and the Methodology of Scientific Research Programmes', pp. 175–6.
[43]Lakatos, 'Introduction: Science and Pseudoscience', pp. 5–6.
[44]Lakatos, 'Falsification and the Methodology of Scientific Research Programmes', p. 176.
[45]Lakatos, 'Introduction: Science and Pseudoscience', p. 6.
[46]Lakatos, 'Falsification and the Methodology of Scientific Research Programmes', p. 137.
[47]Le Grand, *Drifting Continents and Shifting Theories*, Chap. 8.
[48]Takeuchi, Uyeda and Kanamori, *Debate About the Earth*, p. 237.
[49]See Mason, 'A magnetic survey off the west coast of the United States . . .', p. 320.
[50]Takeuchi, Uyeda and Kanamori, *Debate About the Earth*, p. 224.
[51]For another account of the rationale of Lakatos's method of rationally reconstructing the history of science, see his 'History of Science and its Rational Reconstructions'.
[52]Lakatos, 'Falsification and the Methodology of Scientific Research Programmes', p. 155.
[53]Ibid., (his italics).
[54]Ibid., p. 156.
[55]Pannekoek, *A History of Astronomy*, p. 230.
[56]Lakatos, 'Falsification and the Methodology of Scientific Research Programmes', p. 157.
[57]Ibid., p. 158 (his italics).
[58]Lakatos, 'Introduction: Science and Pseudoscience', p. 6.
[59]Feuer, *Einstein and the Generation of Science*, pp. 254–6.
[60]Lakatos, 'Falsification and the Methodology of Scientific Research Programmes', pp. 179–80.
[61]Popper, *Objective Knowledge*, pp. 106, 108–9.
[62]Ibid., pp. 115–16.
[63]Lakatos, 'Falsification and the Methodology of Scientific Research Programmes', p. 176.
[64]Popper, *Objective Knowledge*, pp. 111–12.
[65]A more detailed critique of Popper's notion of objective knowledge is to be found in Bloor, 'Popper's Mystification of Objective Knowledge'. For a more recent discussion, see Parusnikova, 'Popper's World 3 and Human Creativity'.
[66]Lakatos, 'Falsification and the Methodology of Scientific Research Programmes', p. 174.
[67]Musgrave, 'Method or Madness?', p. 475.
[68]Feyerabend, *Against Method*, p. 186 (his italics).
[69]Lakatos, 'History of Science and its Rational Reconstructions', pp. 15–16, note.

4 LAUDAN'S THEORY OF EVOLVING RESEARCH TRADITIONS

[1]Laudan, *Progress and Its Problems*, pp. 5–6 (his italics).
[2]Ibid., p. 14.
[3]Ibid., p. 15.
[4]Ibid.

[5]Ibid., p. 17 (his italics).
[6]Ibid.
[7]Ibid., p. 18.
[8]Ibid., p. 21.
[9]Ibid., pp. 22–3.
[10]Ibid., p. 23.
[11]See pp. 171–6 for further discussion on this issue.
[12]Laudan, *Progress and Its Problems*, pp. 23–4.
[13]Ibid., p. 25.
[14]Ibid., p. 17.
[15]Frankel, 'The Reception and Acceptance of Continental Drift Theory as a Rational Episode in the History of Science', p. 60.
[16]Laudan, *Progress and Its Problems*, p. 28.
[17]Ibid., p. 33.
[18]Ibid., p. 30.
[19]Ibid., p. 34.
[20]Oldroyd, *Darwinian Impacts*, p. 88.
[21]Laudan, *Progress and Its Problems*, p. 35.
[22]Ibid., pp. 35–6.
[23]Ibid., p. 37.
[24]Ibid., pp. 37–8.
[25]Ibid., p. 47.
[26]Ibid., p. 49.
[27]Ibid., pp. 50–1.
[28]Ibid., pp. 51–3.
[29]Ibid., p. 54.
[30]Ibid.
[31]Ibid., p. 55.
[32]Ibid., p. 56.
[33]Takeuchi, Uyeda and Kanamori, *Debate About the Earth*, p. 69.
[34]Laudan, *Progress and Its Problems*, p. 59 (his italics).
[35]Le Grand, *Drifting Continents and Shifting Theories*, p. 50.
[36]Laudan, *Progress and Its Problems*, p. 61.
[37]Dreyer, *A History of Astronomy from Thales to Kepler*, p. 417.
[38]Laudan, *Progress and Its Problems*, pp. 64, 65–6.
[39]Ibid., pp. 78–9 (his italics).
[40]Ibid., p. 81 (his italics).
[41]Ibid., pp. 80–1.
[42]Ibid., p. 81.
[43]Ibid.
[44]Ibid., pp. 96–7.
[45]Ibid., pp. 90, 92.
[46]Ibid., pp. 96–7.
[47]Ibid., p. 98.
[48]Ibid., pp. 98–9 (his italics).
[49]Ibid., p. 99.
[50]Ibid., (his italics).
[51]Frankel, 'The Career of Continental Drift Theory', pp. 58–9, and 'The Reception and Acceptance of Continental Drift Theory as a Rational Episode in the History of Science', p. 53.
[52]Gutting, 'Review of Larry Laudan, Progress and its Problems', p. 93.

[53]Laudan, *Progress and Its Problems*, p. 66.
[54]Ibid., p. 68 (his italics).
[55]Ibid., p. 107.
[56]Ibid., pp. 106–7.
[57]Ibid., p. 107.
[58]Ibid., pp. 107, 108.
[59]Ibid., p. 108.
[60]Ibid., p. 109.
[61]Ibid., p. 112.
[62]Gutting, 'Review of Larry Laudan, Progress and Its Problems', p. 99.
[63]Feyerabend, 'More Clothes from the Emperor's Bargain Basement', p. 66.
[64]Laudan, 'A Problem-Solving Approach to Scientific Progress', p. 150.
[65]Laudan, *Progress and Its Problems*, p. 142.
[66]Ibid., p. 143.
[67]Ibid., p. 144.
[68]Ibid., p. 146.
[69]Ibid., p. 124.
[70]Ibid., p. 132.
[71]Dreyer, *A History of Astronomy from Thales to Kepler*, p. 417.
[72]Cf. Kuhn, *The Copernican Revolution*, p. 204, and Hutchison, 'Planetary Distances as a Test for the Copernican Theory', p. 371.
[73]Laudan, 'A Problem-Solving Approach to Scientific Progress', p. 153.
[74]Laudan, *Progress and Its Problems*, pp. 137, 138.
[75]Laudan, 'A Problem-Solving Approach to Scientific Progress', p. 153.
[76]Feyerabend, 'More Clothes from the Emperor's Bargain Basement', p. 61.
[77]Le Grand, *Drifting Continents and Shifting Theories*, pp. 13–14.

5 THE SOCIOLOGY OF SCIENCE

[1]Merton, *The Sociology of Science*, ed. N. Storer.
[2]Ibid., p. 270.
[3]Ibid., pp. 269, 278.
[4]Storer, *The Social System of Science*, p. 75.
[5]Merton, *The Sociology of Science*, p. 270.
[6]Ibid., p. 269.
[7]Ibid., pp. 270–8.
[8]Barber, *Science and the Social Order*, pp. 86–8.
[9]Merton, *The Sociology of Science*, p. 297.
[10]Ibid., p. 293.
[11]Ball, *A Short Account of the History of Mathematics*, pp. 266, 347–8.
[12]Storer, *The Social System of Science*, p. 83.
[13]Hagstrom, *The Scientific Community*, p. 16.
[14]Storer, *The Social System of Science*, p. 84.
[15]Ibid., p. 85.
[16]Ibid., p. 84.
[17]See Mulkay, *Science and the Sociology of Knowledge*, p. 66.
[18]Mitroff, *The Subjective Side of Science*, pp. 12, 73.
[19]Ibid., p. 79.
[20]Mulkay, *Science and the Sociology of Knowledge*, pp. 68–9.
[21]Ibid., p. 70.

[22]Ibid., p. 72.
[23]Ibid., pp. 70–2. See also Mulkay, 'Interpretation and the Use of Rules'.
[24]Iltis, 'The Leibnizian-Newtonian Debates', p. 372.

6 THE SOCIOLOGY OF SCIENTIFIC KNOWLEDGE

[1]Merton, *The Sociology of Science*, p. 7.
[2]Russell, *The Problems of Philosophy*, p. 110.
[3]Scheffler, *Science and Subjectivity*, p. 19.
[4]Feyerabend, *Against Method*, pp. 27–8.
[5]Collins, 'Son of Seven Sexes', p. 54.
[6]Bloor, *Knowledge and Social Imagery*, pp. 2–3.
[7]Barnes, *Scientific Knowledge and Sociological Theory*, p. 69.
[8]Bloor, *Knowledge and Social Imagery*, pp. 4–5.
[9]Laudan, 'The Pseudo-Science of Science?', pp. 174–5.
[10]Ibid., pp. 181, 183.
[11]Bloor, 'The Strengths of the Strong Programme', p. 206.
[12]Forman, 'Weimar Culture, Causality, and Quantum Theory, 1918–1927', pp. 109–10.
[13]Laudan, 'The Pseudo-Science of Science?', pp. 183–4, 178.
[14]Hesse, *Revolutions and Reconstructions in the Philosophy of Science*, p. 32.
[15]Laudan, 'The Pseudo-Science of Science?', p. 181.
[16]Bloor, *Knowledge and Social Imagery*, pp. 8, 10.
[17]Laudan, 'The Pseudo-Science of Science?', p. 181.
[18]Bloor, *Knowledge and Social Imagery*, p. ix.
[19]Laudan, 'The Pseudo-Science of Science?', pp. 181–2.
[20]Ibid., p. 182.
[21]Ibid.
[22]Ibid., p. 186.
[23]For a more comprehensive introduction to theories of rationality, see H. I. Brown, *Rationality*. See also pp. 176–80.
[24]Laudan, 'The Pseudo-Science of Science?', p. 187
[25]Newton-Smith, *The Rationality of Science*, p. 249.
[26]Clendinnen, 'Epistemic Choice and Sociology', p. 64.
[27]Laudan, 'The Pseudo-Science of Science?', p. 193.
[28]Bloor, 'The Strengths of the Strong Programme', p. 201.
[29]For other arguments against the strong programme not involving cognitive science studies, see J. R. Brown, *The Rational and the Social*, and Nola, 'The Strong Programme for the Sociology of Science, Reflexivity and Relativism'.
[30]See especially Slezak, 'Scientific Discovery by Computer as Empirical Refutation of the Strong Programme'.
[31]Langley, Bradshaw and Simon, 'Rediscovering Physics with BACON.3'; 'Rediscovering Chemistry with the BACON System'.
[32]Slezak, 'Scientific Discovery by Computer', pp. 571–2.
[33]Collins, 'Computers and the Sociology of Scientific Knowledge', pp. 614, 616, 618–19.
[34]Giere, 'Computer Interests and Human Interests', p. 640.
[35]Thagard, 'Welcome to the Cognitive Revolution', p. 655.
[36]Slezak, 'Computers, Contents and Causes', pp. 681–2.
[37]Ibid., p. 685.

[38]Ibid., p. 680.
[39]Further mudslinging between the opposite sides of this debate is to be found in *Social Studies of Science*, 21 (1991): 143–56. Of particular interest is the response to the initial debate on the role of computers in generating scientific theories, by one of the developers of the BACON software, H. A. Simon, 'Comments on the Symposium on "Computer Discovery and the Sociology of Scientific Knowledge"'.
[40]Habermas, *Knowledge and Human Interests.*
[41]Giere, *Explaining Science*, p. 8.
[42]Woolgar, 'Interests and Explanation in the Social Study of Science', pp. 367, 368.
[43]Le Grand, *Drifting Continents and Shifting Theories*, p. 9.
[44]Barnes, *Interests and the Growth of Knowledge*, pp. 37–8.
[45]Giere, *Explaining Science*, p. 53.
[46]Ibid., p. 33.
[47]MacKenzie, *Statistics in Britain 1865–1930.*
[48]Ibid., pp. 217, 221.
[49]Ibid., p. 225.
[50]Barnes and MacKenzie, 'On the Role of Interests in Scientific Change', p. 52.
[51]Ibid., p. 53.
[52]Ibid., p. 61.
[53]Slezak, 'Scientific Discovery by Computer', p. 585 (his italics).
[54]Feyerabend, *Against Method*, pp. 23, 27–8 (his italics).
[55]Ibid., p. 155.
[56]Ibid., p. 79.
[57]Ibid., p. 81 (his italics).
[58]Ibid., p. 77.
[59]Worrall, 'Against Too Much Method', p. 283.
[60]Feyerabend, *Against Method*, p. 98.
[61]Kleiner, 'Feyerabend, Galileo and Darwin', p. 290.
[62]Feyerabend, 'Consolations for the Specialist', p. 209.
[63]Ibid., p. 210.
[64]Feyerabend cites arguments from Benjamin Whorff's *Language, Thought and Reality.*
[65]Feyerabend, *Against Method*, p. 224.
[66]Ibid., p. 225.
[67]Feyerabend, 'Explanation, Reduction, and Empiricism', p. 59.
[68]Feyerabend, *Against Method*, pp. 278–9.
[69]Ibid., p. 296.
[70]Ibid., p. 44.
[71]Ibid., p. 297.
[72]Worrall, 'Against Too Much Method', p. 294.
[73]Latour, *Science in Action*, pp. 2–3.
[74]Ibid., p. 22.
[75]Ibid., p. 23.
[76]Ibid., pp. 23, 24.
[77]Ibid., p. 25 (his italics).
[78]Ibid., p. 29 (his italics).
[79]Cf. Shapin, 'Following Scientists Around', p. 542.
[80]Latour, *Science in Action*, p. 30.

[81]Ibid., pp. 30, 33 (his italics).
[82]Ibid., p. 61.
[83]Ibid., p. 67 (his italics).
[84]Ibid., pp. 74, 75.
[85]Richard Doell quoted in Wood, *The Dark Side of the Earth*, p. 167.
[86]Latour, *Science in Action*, p. 75.
[87]Ibid., p. 84.
[88]Giere, *Explaining Science*, p. 58.

7 RATIONALITY REVISITED AND SOME RELATIONS BETWEEN GENERAL THEORIES AND INDIVIDUAL EXPLANATIONS

[1]Putnam, *Mathematics, Matter and Method*, p. 73.
[2]For example, see Laudan, 'A Confutation of Convergent Realism'.
[3]For a different argument on the role of truth in science, see Cartwright, 'The Truth Doesn't Explain Much'.
[4]*Explaining Science*, p. 4.
[5]*The Scientific Image*, pp. 39–40 (his italics).
[6]See also J. R. Brown, 'Explaining the Success of Science'.
[7]Laudan, 'Explaining the Success of Science'.
[8]Ibid., p. 89 (his italics).
[9]Ibid., pp. 94–5, 97.
[10]Ibid., p. 98.
[11]For another approach to characterizing rationality, see Taylor, 'Rationality'.
[12]*The Rationality of Science*, p. 241.
[13]For example, see ibid., pp. 258ff., and Siegel, 'What is the Question Concerning the Rationality of Science?', who argue for rational aims of science. Giere, *Explaining Science*, p. 10, argues against.
[14]Popper, *Objective Knowledge*, p. 191.
[15]Laudan, *Progress and its Problems*, pp. 130–1 (his italics).
[16]Laudan, *Science and Values*, pp. 62–3.
[17]Le Grand, *Drifting Continents and Shifting Theories*, p. 271; Worrall, 'The Value of a Fixed Methodology'; H. I. Brown, 'Review of Laudan's *Science and Values*'.
[18]The Postscript can be found in the second edition of *The Structure of Scientific Revolutions*, pp. 177–8.
[19]Musgrave, 'Kuhn's Second Thoughts' in Gutting, (ed.), *Paradigms and Revolutions*, pp. 42–3.
[20]Kuhn, Postscript to *The Structure of Scientific Revolutions*, p. 181.
[21]Le Grand, *Drifting Continents and Shifting Theories*, pp. 93–4.
[22]Ibid., pp. 95–6.
[23]'Disagreement in Science', p. 285.
[24]Ibid., p. 290.
[25]Frankel, 'The Reception and Acceptance of Continental Drift Theory as a Rational Episode in the History of Science'; Le Grand, *Drifting Continents and Shifting Theories*; Giere, *Explaining Science*, Chap. 8; Stewart, *Drifting Continents and Colliding Paradigms*.
[26]Takeuchi, Uyeda and Kanamori, *Debate About the Earth*, pp. 69–71.
[27]Frankel, 'The Reception and Acceptance of Continental Drift Theory'.

[28]Le Grand, *Drifting Continents and Shifting Theories*, p. 65.
[29]Laudan, *Progress and its Problems*, pp. 50–1.
[30]Le Grand, *Drifting Continents and Shifting Theories*, p. 65.
[31]Wood, *The Dark Side of the Earth*, p. 83.
[32]Le Grand, *Drifting Continents and Shifting Theories*, p. 66.
[33]Laudan, *Progress and its Problems*, pp. 201–10.
[34]For example, see Jennings, 'Truth, Rationality and the Sociology of Science'.
[35]Further arguments for rationality and against the symmetry tenet may be found in Worrall, 'Rationality, Sociology and the Symmetry Thesis'.

Bibliography and Suggestions for Further Reading

Agassi, J. *Science in Flux. Boston Studies in the Philosophy of Science XXVIII*. Dordrecht: Reidel, 1975.

——— and Jarvie, I. C. (eds). *Rationality: The Critical View*. Dordrecht: Nijhoff, 1987.

Anderson, F. H. (ed.). *The New Organon and Related Writings*. New York: Bobbs-Merrill, 1960.

Arbib, M. A. and Hesse, M. B. *The Construction of Reality*. Cambridge: Cambridge University Press, 1986.

Aristotle. *Prior Analytics*, tr. R. Smith. Indianapolis: Hackett, 1989.

Asquith, P. D. and Giere, R. N. (eds). *PSA 1980: Proceedings of the 1980 Biennial Meeting of the Philosophy of Science Association*. East Lansing, Michigan: Philosophy of Science Association, 1980.

Asquith, P. D. and Hacking, I. (eds). *PSA 1978: Proceedings of the 1978 Biennial Meeting of the Philosophy of Science Association*. East Lansing, Michigan: Philosophy of Science Association, 1978.

Asquith, P. D. and Kyburg, H. E. (eds). *Current Research in Philosophy of Science: Proceedings of the P.S.A. Critical Research Problems Conference*. East Lansing, Michigan: Philosophy of Science Association, 1979.

Asquith, P. D. and Nickles, T. (eds). *PSA 1982: Proceedings of the 1982 Biennial Meeting of the Philosophy of Science Association*. East Lansing, Michigan: Philosophy of Science Association, 1982.

Ayer, A. J. *The Problem of Knowledge*. Harmondsworth: Penguin, 1956.

———. *Language, Truth and Logic*. Harmondsworth: Penguin, 1958.

Bacon, F. *Novum Organum*, ed. T. Fowler. Oxford: Clarendon Press, 1878.

Ball, W. W. R. *A Short Account of the History of Mathematics*. New York: Dover, 1960.

Barber, B. *Science and the Social Order*. New York: Collier, 1962.

———. 'On the Relations between Philosophy of Science and Sociology of Science' in P. D. Asquith and H. E. Kyburg (eds), *Current Research in Philosophy of Science: Proceedings of the P.S.A. Critical Research Problems Conference*. East Lansing, Michigan: Philosophy of Science Association, 1979.

Barnes, S. B. (ed.). *Sociology of Science*. Harmondsworth: Penguin, 1972.

———. *Scientific Knowledge and Sociological Theory*. London: Routledge & Kegan Paul, 1974.

———. 'Natural Rationality: A Neglected Concept in the Social Sciences'. *Philosophy of the Social Sciences*, 6 (1976): 115–26.

———. *Interests and the Growth of Knowledge*. London: Routledge & Kegan Paul, 1977.

———. *T. S. Kuhn and Social Science*. London: Macmillan, 1982.

———. *About Science*. Oxford: Blackwell, 1985.

——— and Bloor, D. 'Relativism, Rationalism and the Sociology of Knowledge' in M. Hollis and S. Lukes (eds), *Rationality and Relativism*. Oxford: Blackwell, 1982.

——— and Edge, D. (eds). *Science in Context: Readings in the Sociology of Science*. Cambridge, Massachusetts: MIT Press, 1982.

——— and MacKenzie, D. 'On the Role of Interests in Scientific Change' in R. Wallis (ed.), *On the Margins of Science: The Social Construction of Rejected Knowledge*. Keele: University of Keele, 1979.

——— and Shapin, S. (eds). *Natural Order: Historical Studies of Scientific Culture*. Beverly Hills: Sage, 1979.

Biagioli, M. 'The Anthropology of Incommensurability'. *Studies in History and Philosophy of Science*, 21 (1990): 183–209.

Blake, R. M., Ducasse, C. J. and Madden, E. H. *Theories of*

Scientific Method: The Renaissance through the Nineteenth Century. Seattle: University of Washington Press, 1960.

Bloor, D. 'Popper's Mystification of Objective Knowledge'. *Science Studies*, 4 (1974): 65–76.

———. *Science and Social Imagery*. London: Routledge & Kegan Paul, 1976.

———. 'The Sociology of Reasons: Or Why "Epistemic Factors" are Really "Social Factors"' in J. R. Brown (ed.), *Scientific Rationality: The Sociological Turn*. Dordrecht: Reidel, 1984.

———. 'The Strengths of the Strong Programme'. *Philosophy of the Social Sciences*, 11 (1981): 199–213. Reprinted in J. R. Brown (ed.), *Scientific Rationality: The Sociological Turn*. Dordrecht: Reidel, 1984.

Bradshaw, G. F., Langley, P. W. and Simon, H. A. 'Studying Scientific Discovery by Computer Simulation'. *Science*, 222 (1983): 971–5.

Brannigan, A. *The Social Basis of Scientific Discoveries*. Cambridge: Cambridge University Press, 1981.

Brown, H. I. 'Review of *Science and Values*'. *Philosophical Review*, 95 (1986): 439–41.

———. *Rationality*. London: Routledge & Kegan Paul, 1988.

Brown, J. R. (ed.). *Scientific Rationality: The Sociological Turn*. Dordrecht: Reidel, 1984.

———. 'Explaining the Success of Science'. *Ratio*, 27 (1985): 49–66.

———. *The Rational and the Social*. London: Routledge, 1989.

Buchdahl, G. 'History of Science and Criteria of Choice' in R. H. Stuewer (ed.). *Historical and Philosophical Perspectives of Science. Minnesota Studies in the Philosophy of Science V*. Minneapolis: University of Minnesota Press, 1970.

Buck, R. C. and Cohen, R. S. (eds). *PSA 1970: In Memory of Rudolf Carnap*. Dordrecht: Reidel, 1971.

Bukharin, N. I. *et al. Science at the Crossroads*. London: Khiga, 1931. Also published by New Left Books, London, and by Cass Publications, London.

Bunge, M. *The Myth of Simplicity: Problems of Scientific Philosophy*. New Jersey: Prentice-Hall, 1963.

Butts, R. E. 'Scientific Progress: The Laudan Manifesto'. *Philosophy of the Social Sciences,* 9 (1979): 475–83.

Carazzi, A. 'The Reaction in Continental Europe to Wegener's Theory of Continental Drift'. *Earth Sciences History,* 4 (1985): 122–37.

Cartwright, N. 'The Truth Doesn't Explain Much'. *American Philosophical Quarterly,* 17 (1980): 159–63. Reprinted in her *How the Laws of Physics Lie.* Oxford: Clarendon, 1983.

Chalmers, A. F. 'An Improvement and a Critique of Lakatos's Methodology of Scientific Research Programmes'. *Methodology and Science,* 13 (1980): 2–27.

———. *What Is This Thing Called Science? An Assessment of the Nature and Status of Science and its Methods.* Milton Keynes: Open University Press, 1982.

Clendinnen, F. J. 'Rational Expectation and Simplicity' in R. McLaughlin (ed.). *What? Where? When? Why? Essays on Induction, Space and Time, Explanation.* Dordrecht: Reidel, 1982.

———. 'The Rationality of Method Verus Historical Relativism'. *Studies in History and Philosophy of Science,* 14 (1983): 23–38.

———. 'Epistemic Choice and Sociology'. *Metascience,* 1/2 (1984): 61–9.

———. 'Realism and Underdetermination of Theory'. *Synthese,* 81 (1989): 63–90.

Cohen, R. S., Feyerabend, P. K. and Wartofsky, M. W. (eds). *Essays in Memory of Imre Lakatos. Boston Studies in the Philosophy of Science IXL.* Dordrecht: Reidel, 1976.

Collins, H. M. 'Son of Seven Sexes: The Social Destruction of a Physical Phenomenon'. *Social Studies of Science,* 11 (1981): 33–62.

——— (ed.). *Sociology of Scientific Knowledge: A Sourcebook.* Bath: Bath University Press, 1982.

———. 'Computers and the Sociology of Scientific Knowledge'. *Social Studies of Science,* 19 (1989): 613–24.

Colodny, R. (ed.). *Beyond the Edge of Certainty.* New Jersey, Prentice-Hall, 1965.

———. *Mind and Cosmos: Essays in Contemporary Science and Philosophy*. Pittsburgh: Pittsburgh University Press, 1966.

———. *The Nature and Function of Scientific Theories*. Pittsburgh: Pittsburgh University Press, 1970.

Cranston, M. 'Francis Bacon' in P. Edwards (ed.), *The Encyclopedia of Philosophy*. New York: Macmillan, 1967.

Cushing, J. T., Delaney, C. F. and Gutting, G. M. (eds). *Science and Reality: Recent Work in the Philosophy of Science*. Notre Dame, Indiana: Notre Dame University Press, 1984.

Dilworth, C. *Scientific Progress: A Study Concerning the Nature of the Relation Between Successive Theories*. Dordrecht: Reidel, 1986.

Dolby, R. G. A. 'Sociology of Knowledge in Natural Science'. *Science Studies*, 1 (1971): 3–21. Reprinted (in part) in B. Barnes (ed.), *Sociology of Science*. Harmondsworth: Penguin, 1972.

Donovan, A., Laudan, L. and Laudan, R. (eds). *Scrutinizing Science: Empirical Studies of Scientific Change*. Dordrecht: Kluwer, 1988.

Doppelt, G. 'Kuhn's Epistemological Relativism: An Interpretation and Defense'. *Inquiry*, 21 (1978): 33–86. Reprinted in J. W. Meiland and M. Krausz (eds), *Relativism: Cognitive and Moral*. Notre Dame, Indiana: Notre Dame University Press, 1982.

———. 'Relativism and Recent Pragmatic Conceptions of Scientific Rationality' in N. Rescher (ed.), *Scientific Explanation and Understanding*. Lanham, Maryland: University Press of America, 1983.

Drake, S. *Galileo*. Oxford: Oxford University Press, 1980.

Dreyer, J. L. E. *A History of Astronomy from Thales to Kepler*, revised by W. H. Stahl. New York: Dover, 1953.

Edwards, P. (ed.). *The Encylopedia of Philosophy*. New York: Macmillan, 1967.

Elkana, Y. (ed.). *The Interaction between Science and Philosophy*. Atlantic Highlands, New Jersey: Humanities Press, 1974.

Ezrahi, Y. 'The Political Resources of Science' in B. Barnes (ed.), *Sociology of Science*. Harmondsworth: Penguin, 1972. Reprinted from *Science Studies*, 1 (1971): 117–33.

Feuer, L. *Einstein and the Generation of Science.* New York: Basic Books, 1982.

Feyerabend, P. K. 'Explanation, Reduction, and Empiricism' in H. Fiegl and G. Maxwell (eds). *Scientific Explanation, Space, and Time. Minnesota Studies in the Philosophy of Science III.* Minneapolis: University of Minnesota Press, 1962. Reprinted in P. K. Feyerabend, *Realism, Rationalism and Scientific Method. Philosophical Papers Volume 1.* Cambridge: Cambridge University Press, 1981.

———. 'Problems of Empiricism' in R. Colodny (ed.). *Beyond the Edge of Certainty.* New Jersey: Prentice-Hall, 1965.

———. 'Consolations for the Specialist' in I. Lakatos and A. Musgrave (eds), *Criticism and the Growth of Knowledge.* Cambridge: Cambridge University Press, 1970. Reprinted in P. K. Feyerabend, *Problems of Empiricism. Philosophical Papers Volume 2.* Cambridge: Cambridge University Press, 1981.

———. 'Philosophy of Science: A Subject with a Great Past' in R. H. Stuewer (ed.). *Historical and Philosophical Perspectives of Science. Minnesota Studies in the Philosophy of Science V.* Minneapolis: University of Minnesota Press, 1970.

———. 'Problems of Empiricism Part II' in R. Colodny (ed.). *The Nature and Function of Scientific Theories.* Pittsburgh: Pittsburgh University Press, 1970.

———. *Against Method: Outline of an Anarchistic Theory of Knowledge.* London: New Left Books, 1975.

———. 'On the Critique of Scientific Reason' in R. S. Cohen, P. K. Feyerabend, and M. W. Wartofsky (eds), *Essays in Memory of Imre Lakatos. Boston Studies in the Philosophy of Science IXL.* Dordrecht: Reidel, 1976. Reprinted in C. Howson (ed.), *Method and Appraisal in the Physical Sciences.* Cambridge: Cambridge University Press, 1976.

———. *Science in a Free Society.* London: New Left Books, 1978.

———. 'More Clothes from the Emperor's Bargain Basement (Review of L. Laudan: *Progress and its Problems*)'. *British Journal for the Philosophy of Science,* 32 (1981): 57–71. Reprinted in P. K. Feyerabend, *Problems of Empiricism. Philosophical*

Papers Volume 2. Cambridge: Cambridge University Press, 1981.

———. *Problems of Empiricism. Philosophical Papers Volume 2*. Cambridge: Cambridge University Press, 1981.

———. *Realism, Rationalism and Scientific Method. Philosophical Papers Volume 1*. Cambridge: Cambridge University Press, 1981.

———. *Farewell to Reason*. London: Verso, 1987.

Fine, A. and Leplin, J. (eds). *PSA 1988: Proceedings of the 1988 Biennial Meeting of the Philosophy of Science Association, Volume One*. East Lansing, Michigan: Philosophy of Science Association, 1988.

Forman, P. 'Weimar Culture, Causality, and Quantum Theory, 1918–1927: Adaptation by German Physicists and Mathematicians to a Hostile Intellectual Environment' in R. McCormmach (ed.), *Historical Studies in the Physical Sciences Volume 3*. Philadelphia: University of Pennsylvania Press, 1971.

Foucault, M. *The Archaeology of Knowledge*. London: Tavistock, 1974.

Frankel, H. 'The Career of Continental Drift Theory'. *Studies in History and Philosophy of Science*, 10 (1979): 21–66.

———. 'The Reception and Acceptance of Continental Drift Theory as a Rational Episode in the History of Science' in S. H. Mauskopf (ed.), *The Reception of Unconventional Science*. Boulder, Colorado: Westview Press, 1979.

Freudenthal, G. 'How Strong is Dr. Bloor's "Strong Programme"'. *Studies in History and Philosophy of Science*, 10 (1979): 67–83.

Fuller, S. *Philosophy of Science and Its Discontents*. Boulder, Colorado: Westview Press, 1989.

———, DeMay, M., Shinn, T. and Woolgar, S. (eds). *The Cognitive Turn: Sociological and Psychological Perspectives on Science*. Dordrecht: Kluwer, 1989.

Giere, R. 'Philosophy of Science Naturalized'. *Philosophy of Science*, 52 (1985): 331–56.

———. *Explaining Science: A Cognitive Approach*. Chicago: Chicago University Press, 1988.

———. 'Computer Interests and Human Interests'. *Social Studies of Science*, 19 (1989): 638–43.

———. 'Scientific Rationality as Instrumental Rationality'. *Studies in History and Philosophy of Science*, 20 (1989): 377–84.

Gieryn, T. F. (ed.). *Science and Social Structure: A Festschrift for Robert K. Merton. Transactions of the New York Academy of Science*, series II, vol. 39, 1980.

Glymour, C. *Theory and Evidence*. Princeton, New Jersey: Princeton University Press, 1980.

Gregg, J. R. and Harris, F. T. C. (eds). *Form and Strategy in Science*. Dordrecht: Reidel, 1964.

Gregory, R. L. *Eye and Brain: The Psychology of Seeing*. London: Weidenfeld and Nicolson, 1977.

Grobler, A. 'Between Rationalism and Relativism. On Larry Laudan's Model of Scientific Rationality'. *British Journal for the Philosophy of Science*, 41 (1990): 493–507.

Grosser, M. *The Discovery of Neptune*. Cambridge, Massachusetts: Harvard University Press, 1962.

Gutting, G. (ed.). *Paradigms and Revolutions: Applications and Appraisals of Thomas Kuhn's Philosophy of Science*. Notre Dame, Indiana: Notre Dame University Press, 1980.

———. 'Review of Larry Laudan, *Progress and its Problems* . . .'. *Erkenntnis*, 15 (1980): 91–103.

Habermas, J. *Knowledge and Human Interests*, tr. J. J. Shapiro. London: Heinemann, 1972. Original German edn 1968.

Hacking, I. 'Lakatos's Philosophy of Science'. *British Journal for the Philosophy of Science*, 30 (1979): 381–410. Reprinted in I. Hacking (ed.). *Scientific Revolutions*. Oxford: Oxford University Press, 1981.

——— (ed.). *Scientific Revolutions*. Oxford: Oxford University Press, 1981.

Hagstrom, W. O. *The Scientific Community*. New York: Basic Books, 1965.

Hanson, N. R. *Perception and Discovery: An Introduction to Scientific Inquiry*. San Francisco: Freeman, Cooper & Co., 1969.

Harre, R. M. *The Philosophies of Science: An Introductory Survey.* Oxford: Oxford University Press, 1972.

Hempel, C. G. *Philosophy of Natural Science.* New Jersey: Prentice-Hall, 1966.

———. *Aspects of Scientific Explanation and Other Essays in the Philosophy of Science.* New York: Free Press, 1970.

Hesse, M. B. 'Is There an Independent Observation Language?' in R. Colodny (ed.), *The Nature and Function of Scientific Theories.* Pittsburgh: Pittsburgh University Press, 1970.

———. *The Structure of Scientific Inference.* London: Macmillan, 1974.

———. *Revolutions and Reconstructions in the Philosophy of Science.* Sussex: Harvester, 1980.

Hessen, B. 'The Social and Economic Roots of Newton's Principia' in N. I. Bukharin *et al., Science at the Crossroads.* London: Khiga, 1931. Also published by New Left Books, London, and Cass, London.

Hilpinen, R. (ed.). *Rationality in Science: Studies in the Foundations of Science and Ethics.* Dordrecht: Reidel, 1980.

Hollis, M. 'The Social Destruction of Reality' in M. Hollis and S. Lukes (eds), *Rationality and Relativism.* Oxford: Blackwell, 1982.

——— and Lukes, S. (eds). *Rationality and Relativism.* Oxford: Blackwell, 1982.

Home, R. W. 'Franklin's Electrical Atmospheres'. *British Journal for the History of Science,* 6 (1972–73): 131–51.

Howson, C. (ed.). *Method and Appraisal in the Physical Sciences.* Cambridge: Cambridge University Press, 1976.

Hunt, E. K. and Sherman, H. H. *Economics: An Introduction to Traditional and Radical Views.* New York: Harper & Row, 1978.

Hutchison, K. R. 'Planetary Distances as a Test for the Copernican Theory'. *British Journal for the Philosophy of Science,* 34 (1983): 369–72.

Iltis, C. 'Leibniz and the Vis-Viva Controversy'. *Isis*, 62 (1971): 21–35.

———. 'The Leibnizian-Newtonian Debates: Natural Philosophy and Social Psychology'. *British Journal for the History of Science*, 6 (1972–1973): 343–77.

Jennings, R. C. 'Truth, Rationality and the Sociology of Science'. *British Journal for the Philosophy of Science*, 35 (1984): 201–11.

Jones, B. 'Plate Tectonics: a Kuhnian Case?'. *New Scientist*, 29 August 1974.

Jones, K. 'Is Kuhn a Sociologist?'. *British Journal for the Philosophy of Science*, 37 (1986): 443–52.

Kant, I. *The Critique of Pure Reason*, tr. J. M. D. Meiklejohn. London and New York: George Bell and Sons, 1893.

Keynes, J. M. *The General Theory of Employment, Interest and Money*. New York: Harcourt Brace Jovanovich, 1936.

Kleiner, S. A. 'Feyerabend, Galileo and Darwin'. *Studies in History and Philosophy of Science*, 10 (1979): 285–309.

Knorr-Cetina, K. D. *The Manufacture of Knowledge: An Essay on the Constructivist and Contextual Nature of Science*. Oxford: Pergamon, 1981.

——— and Mulkay, M. (eds). *Science Observed: Perspectives on the Social Study of Science*. London: Sage, 1983.

Koertge, N. 'The Problem of Appraising Scientific Theories' in P. D. Asquith and H. E. Kyburg (eds), *Current Research in Philosophy of Science: Proceedings of the P.S.A. Critical Research Problems Conference*. East Lansing, Michigan: Philosophy of Science Association, 1979.

———. 'In Praise of Truth and Substantive Rationality: Comments on Laudan's *Progress and its Problems*' in P. D. Asquith and I. Hacking (eds), *PSA 1978: Proceedings of the 1978 Biennial Meeting of the Philosophy of Science Association*, vol. 2. East Lansing, Michigan: Philosophy of Science Association, 1981.

Krige, J. *Science, Revolution and Discontinuity*. Sussex: Harvester, 1980.

Kuhn, T. S. *The Structure of Scientific Revolutions*, 2nd edn. Chicago: Chicago University Press, 1970. Original edn 1962.

———. 'Reflections on my Critics' in I. Lakatos and A. Musgrave (eds), *Criticism and the Growth of Knowledge*. Cambridge: Cambridge University Press, 1970.

———. 'Notes on Lakatos' in R. C. Buck and R. S. Cohen (eds), *PSA 1970: In Memory of Rudolf Carnap*. Dordrecht: Reidel, 1971.

———. *The Copernican Revolution*. Cambridge, Massachusetts: Harvard University Press, 1974.

———. *The Essential Tension: Selected Studies in Scientific Tradition and Change*. Chicago: Chicago University Press, 1977.

———. 'Second Thoughts on Paradigms' in F. Suppe (ed.), *The Structure of Scientific Theories*. Urbana, Illinois: University of Illinois Press, 1977.

———. 'Metaphor in Science' in A. Ortony (ed.). *Metaphor and Thought*. Cambridge: Cambridge University Press, 1979.

———. 'Commensurability, Comparability, Communicability' in *PSA 1982 Proceedings of the 1982 Biennial Meeting of the Philosophy of Science Association*, vol. 2. East Lansing, Michigan: Philosophy of Science Association, 1982.

———. 'Rationality and Theory Choice'. *Journal of Philosophy*, 80 (1983): 563–71.

Lakatos, I. 'Criticism and the Methodology of Scientific Research Programmes'. *Proceedings of the Aristotelian Society*, new series, 69 (1968–69): 149–86.

———. 'Falsification and the Methodology of Scientific Research Programmes' in I. Lakatos and A. Musgrave (eds), *Criticism and the Growth of Knowledge*. Cambridge: Cambridge University Press, 1970. Reprinted in J. Worrall and G. Currie (eds), *Imre Lakatos: Philosophical Papers Volume 1*. Cambridge: Cambridge University Press, 1978.

———. 'History of Science and Its Rational Reconstructions' in R. C. Buck and R. S. Cohen (eds). *PSA 1970: In Memory of Rudolf Carnap* Dordrecht: Reidel, 1971. Reprinted in Y. Elkana (ed.), *The Interaction Between Science and Philosophy*. Atlantic Highlands, New Jersey: Humanities Press, 1974; in C. Howson (ed.), *Method and Appraisal in the Physical Sciences*. Cambridge: Cambridge University Press, 1976; in J. Worrall and G. Currie (eds), *Imre Lakatos: Philosophical*

Papers Volume 1. Cambridge: Cambridge University Press, 1978, and in I. Hacking (ed.), *Scientific Revolutions.* Oxford: Oxford University Press, 1981.
———. 'Introduction: Science and Pseudoscience' in J. Worrall and G. Currie (eds). *Imre Lakatos: Philosophical Papers Volume 1.* Cambridge: Cambridge University Press, 1978.
——— and Musgrave, A. (eds). *Problems in the Philosophy of Science.* Amsterdam: North-Holland Publishers, 1968.
——— and Musgrave, A. (eds). *Criticism and the Growth of Knowledge.* Cambridge: Cambridge University Press, 1970.
Langley, P., Bradshaw, G. L. and Simon, H. A. 'Rediscovering Physics with BACON.3'. *Proceedings of the Sixth International Joint Conference on Artificial Intelligence,* 1 (Tokyo, August 1979): 505–7.
———. 'Rediscovering Chemistry with the BACON System' in R. S. Michalski, J. G. Carbonell and T. M. Mitchell (eds), *Machine Learning: An Artificial Intelligence Approach.* New York: Springer, 1980.
Latour, B. *Science in Action: How to Follow Scientists and Engineers Through Society.* Milton Keynes: Open University Press, 1987.
———. 'Clothing the Naked Truth' in H. Lawson and L. Appignanesi, (eds), *Dismantling Truth: Reality in the Post-Modern World.* London: Weidenfeld and Nicolson, 1989.
——— and Woolgar, S. *Laboratory Life: The Social Construction of Scientific Facts.* Beverly Hills: Sage, 1978.
Laudan, L. 'The *Vis viva* Controversy, A Post-Mortem'. *Isis,* 59 (1968): 131–43.
———. *Progress and Its Problems: Towards a Theory of Scientific Growth.* Berkeley and Los Angeles: University of California Press, 1977.
———. 'The Philosophy of Progress . . .' in P. D. Asquith and I. Hacking (eds). *PSA 1978: Proceedings of the 1978 Biennial Meeting of the Philosophy of Science Association,* vol. 2. East Lansing, Michigan: Philosophy of Science Association, 1978.
———. 'Views of Progress: Separating the Pilgims from the Rakes'. *Philosophy of the Social Sciences,* 10 (1980): 273–86.

———. 'Problems, Truth, and Consistency'. *Studies in History and Philosophy of Science*, 13 (1981): 73–80.

———. 'A Confutation of Convergent Realism'. *Philosophy of Science*, 48 (1981): 19–49. Reprinted in J. Leplin (ed.), *Scientific Realism*. Berkeley and Los Angeles: University of California Press, 1984.

———. 'A Problem-Solving Approach to Scientific Progress' in I. Hacking (ed.), *Scientific Revolutions*. Oxford: Oxford University Press, 1981.

———. 'The Pseudo-Science of Science?'. *Philosophy of the Social Sciences*, 11 (1981): 173–98. Reprinted in J. R. Brown (ed.), *Scientific Rationality: The Sociological Turn*. Dordrecht: Reidel, 1984.

———. 'A Note on Collins's Blend of Relativism and Empiricism'. *Social Studies of Science*, 12 (1982): 131–32.

———. 'Explaining the Success of Science: Beyond Epistemic Realism and Relativism' in J. T. Cushing, C. F. Delaney, and G. M. Gutting. *Science and Reality: Recent Work in the Philosophy of Science*. Notre Dame, Indiana: Notre Dame University Press, 1984.

———. *Science and Values: The Aims of Science and Their Role in Scientific Debate*. Berkeley and Los Angeles: University of California Press, 1984.

———. 'Progress or Rationality? The Prospects for a Normative Naturalism'. *American Philosophical Quarterly*, 24 (1987): 19–31.

———. 'Relativism, Naturalism and Reticulation'. *Synthese*, 71 (1987): 221–34.

———. 'If it ain't Broke, Don't fix it'. *British Journal for the Philosophy of Science*, 40 (1989): 369–75.

———. 'Normative Naturalism'. *Philosophy of Science*, 57 (1990): 44–59.

———. *Science and Relativism*. Chicago: Chicago University Press, 1990.

——— *et al.* 'Scientific Change: Philosophical Models and Historical Research'. *Synthese*, 69 (1986): 141–223.

Laudan, R. 'The Recent Revolution in Geology and Kuhn's Theory of Scientific Change' in P. D. Asquith and I. Hacking (eds), *PSA 1978: Proceedings of the 1978 Biennial Meeting of the Philosophy of Science Association*, vol. 2. East Lansing, Michigan: Philosophy of Science Association, 1981. Reprinted in G. Gutting (ed.), *Paradigms and Revolutions: Applications and Appraisals of Thomas Kuhn's Philosophy of Science*. Notre Dame, Indiana: Notre Dame University Press, 1980.

——— (ed.). *The Nature of Technological Knowledge: Are Models of Scientific Change Relevant?* Boston: Reidel, 1984.

——— and Laudan, L. 'Dominance and the Disunity of Method: Solving the Problems of Innovation and Consensus'. *Philosophy of Science*, 56 (1989): 221–37.

Le Grand, H. E. 'Specialities, Problems, and Localism: The Reception of Continental Drift in Australia 1920–1940'. *Earth Sciences History*, 5 (1987): 84–95.

———. *Drifting Continents and Shifting Theories*. Cambridge: Cambridge University Press, 1988.

Leicester, H. M. *The Historical Background of Chemistry*. New York: Dover, 1971.

Lemmon, E. J. *Beginning Logic*. London: Nelson, 1965.

Leplin, J. (ed.). *Scientific Realism*. Berkeley and Los Angeles: University of California Press, 1984.

Levin, J. 'Must Reasons be Rational?'. *Philosophy of Science*, 55 (1988): 199–217.

Losee, J. 'Laudan on Progress in Science'. *Studies in History and Philosophy of Science*, 9 (1978): 333–40.

———. *A Historical Introduction to the Philosophy of Science*, 2nd edn. Oxford: Oxford University Press, 1980.

———. *Philosophy of Science and Historical Enquiry*. Oxford: Oxford University Press, 1987.

Lugg, A. 'Disagreement in Science'. *Zeitschrift für allgemeine Wissenschaftstheorie*, IX/2 (1978): 276–92.

———. 'Overdetermined Problems in Science'. *Studies in History and Philosophy of Science*, 9 (1978): 1–18.

———. 'An Alternative to the Traditional Model? Laudan on Disagreement and Consensus in Science'. *Philosophy of Science*, 53 (1986): 419–24.

———. 'Critical Notice of Paul Feyerabend, *Farewell to Reason*'. *Canadian Journal of Philosophy*, 20 (1991): 109–20.

McCormmach, R. (ed.). *Historical Studies in the Physical Sciences Volume 3*. Philadelphia: University of Pennsylvania Press, 1971.

MacKenzie, D. A. *Statistics in Britain 1865–1930: The Social Construction of Scientific Knowledge*. Edinburgh: Edinburgh University Press, 1981.

McLaughlin, R. (ed.). *What? Where? When? Why? Essays on Induction, Space and Time, Explanation*. Dordrecht: Reidel, 1982.

McMullin, E. (ed.). *Construction and Constraint*. Notre Dame, Indiana: Notre Dame University Press, 1988.

Magee, B. *Popper*. London: Fontana, 1973.

Marinov, M. 'Cognitive Values and Scientific Rationality'. *International Studies in the Philosophy of Science*, 1 (1987): 223–32.

Mason, R. 'A magnetic survey off the west coast of the United States between lat. 32° & 26° N, long. 121° & 128° W.'. *Geophysics Journal*, 1 (1958): 320.

Masterman, M. 'The Nature of Paradigm' in I. Lakatos, and A. Musgrave (eds), *Criticism and the Growth of Knowledge*. Cambridge: Cambridge University Press, 1970.

Meiland, J. W. 'Kuhn, Scheffler, and Objectivity in Science'. *Philosophy of Science*, 41 (1974): 179–87.

——— and Krausz, M. (eds). *Relativism: Cognitive and Moral*. Notre Dame, Indiana: Notre Dame University Press, 1982.

Merton, R. K. *The Sociology of Science: Theoretical and Empirical Investigations*, ed. N. Storer. Chicago: Chicago University Press, 1973.

Michalski, R. S., Carbonell, J. G. and Mitchell, T. M. (eds). *Machine Learning: An Artificial Intelligence Approach*. New York: Springer, 1980.

Miller, R. W. *Fact and Method: Explanation, Confirmation and Reality in the Natural and Social Sciences*. Princeton, New Jersey: Princeton University Press, 1987.

Mitroff, I. I. *The Subjective Side of Science: A Philosophical Inquiry into the Psychology of the Apollo Moon Scientists*. Amsterdam: Elsevier, 1974.

Mokrzycki, E. *Philosophy of Science and Sociology: From the Methodological Doctrine to Research Practice.* London: Routledge & Kegan Paul, 1983.

Mulkay, M. *Science and the Sociology of Knowledge.* London and Boston: Allen & Unwin, 1979.

———. 'Interpretation and the Use of Rules: The Case of the Norms of Science' in T. F. Gieryn (ed.), *Science and Social Structure: A Festschrift for Robert K. Merton. Transactions of the New York Academy of Science*, series II, vol. 39, 1980.

———. *Sociology of Science: A Sociological Pilgrimage.* Milton Keynes: Open University Press, 1991.

——— and Gilbert, G. N. 'Accounting for Error: How Scientists Construct their Social World When they Account for Correct and Incorrect Belief'. *Sociology*, 16 (1982): 165–83.

Musgrave, A. E. 'Kuhn's Second Thoughts'. *British Journal for the Philosophy of Science*, 22 (1971): 287–306. Reprinted in G. Gutting (ed.), *Paradigms and Revolutions: Applications and Appraisals of Thomas Kuhn's Philosophy of Science.* Notre Dame, Indiana: Notre Dame University Press, 1980.

———. 'Method or Madness? Can the Methodology of Research Programmes Be Rescued From Epistemological Anarchism?' in R. S. Cohen, P. K. Feyerabend and M. W. Wartofsky (eds), *Essays in Memory of Imre Lakatos. Boston Studies in the Philosophy of Science* IXL. Dordrecht: Reidel, 1976.

———. 'Why did oxygen supplant phlogiston? Research programmes in the Chemical Revolution' in C. Howson (ed.), *Method and Appraisal in the Physical Sciences.* Cambridge: Cambridge University Press, 1976.

———. 'The Ultimate Argument for Scientific Realism' in R. Nola (ed.), *Relativism and Realism in Science.* Dordrecht: Kluwer, 1988.

Nagel, E. *The Structure of Science.* London: Routledge & Kegan Paul, 1961.

———. *Teleology Revisited.* New York: Columbia University Press, 1979.

Newton-Smith, W. H. 'The Underdetermination of Theory by Data'. *Proceedings of the Aristotelian Society* (Supplementary

Volume), 52 (1978): 71–91. Reprinted in R. Hilpinen (ed.), *Rationality in Science: Studies in the Foundations of Science and Ethics*. Dordrecht: Reidel, 1980.

———. *The Rationality of Science*. London: Routledge & Kegan Paul, 1981.

Nickles, T. *Scientific Discovery: Case Studies. Boston Studies in the Philosophy of Science LX*. Dordrecht: Reidel, 1980.

——— (ed.). *Scientific Discovery, Logic and Rationality. Boston Studies in the Philosophy of Science LVI*. Dordrecht: Reidel, 1980.

Nola, R. (ed.). *Relativism and Realism in Science*. Dordrecht: Kluwer, 1988.

———. 'The Strong Programme for the Sociology of Science, Reflexivity and Relativism'. *Inquiry*, 33 (1990): 273–96.

O'Hear, A. *Karl Popper*. London: Routledge & Kegan Paul, 1980.

———. *Introduction to Philosophy of Science*. Oxford: Oxford University Press, 1989.

Oldroyd, D. R. *Darwinian Impacts: an Introduction to the Darwinian Revolution*. Kensington, N.S.W.: New South Wales University Press, 1980.

———. *The Arch of Knowledge: An Introductory Study of the History of the Philosophy and Methodology of Science*. Kensington, N.S.W.: New South Wales University Press, 1986.

Ortony, A. (ed.). *Metaphor and Thought*. Cambridge: Cambridge University Press, 1979.

Orwell, G. *Nineteen Eighty-Four*. Oxford: Clarendon, 1984. First edn England: Secker & Warburg, 1949.

Pannekoek, A. *A History of Astronomy*. London: Allen & Unwin, 1961.

Parusnikova, Z. 'Popper's World 3 and Human Creativity'. *International Studies in the Philosophy of Science*, 4 (1990): 263–9.

Passmore, J. 'Logical Positivism' in P. Edwards (ed.), *The Encyclopedia of Philosophy*. New York: Macmillan, 1967.

Pettit, P. 'The Strong Sociology of Knowledge Without Relativism' in R. Nola (ed.), *Relativism and Realism in Science*. Dordrecht: Kluwer, 1988.

Pias, A. *'Subtle is the Lord . . . ': The Science and the Life of Albert Einstein.* Oxford: Oxford University Press, 1984.

Pickering, A. *Constructing Quarks: A Sociological History of Particle Physics.* Edinburgh: Blackwell, 1984.

Pitt, J. (ed.). *Change and Progress in Modern Science.* Dordrecht: Reidel, 1980.

Polanyi, M. *Personal Knowledge.* Chicago: Chicago University Press, 1958.

———. *The Tacit Dimension.* London: Routledge & Kegan Paul, 1967.

Popper, K. R. *The Logic of Scientific Discovery.* London: Hutchinson, 1975. Translated from the original 1934 German edn.

———. *Conjectures and Refutations: The Growth of Scientific Knowledge*, 4th edn (revised). London: Routledge & Kegan Paul, 1972. Original edn 1963.

———. *Objective Knowledge: An Evolutionary Approach.* Oxford: Clarendon, 1972.

———. *Realism and the Aim of Science.* London: Hutchinson, 1982.

Putnam, H. *Mathematics, Matter and Method. Philosophical Papers Volume 1.* Cambridge: Cambridge University Press, 1975.

Quine, W. V. O. *From a Logical Point of View.* Cambridge, Massachusetts: Harvard University Press, 1952.

———. *Word and Object.* Cambridge, Massachusetts: MIT Press, 1960.

———. 'On Simple Theories of a Complex World'. *Synthese*, 15 (1963): 103–6. Reprinted in J. R. Gregg and F. T. C. Harris (eds), *Form and Strategy in Science.* Dordrecht: Reidel, 1964, and in W. V. O. Quine, *The Ways of Paradox and Other Essays.* Cambridge, Massachusetts: Harvard University Press, 1976.

——— and Ullian, J. S. *The Web of Belief.* New York: Random House, 1970.

Reisch, G. A. 'Did Kuhn Kill Logical Empiricism?'. *Philosophy of Science*, 58 (1991): 264–77.

Rescher, N. (ed.). *Scientific Explanation and Understanding.* Lanham, Maryland: University Press of America, 1983.

Richards, S. *Philosophy and Sociology of Science: An Introduction.* Oxford: Blackwell, 1983.
Rossi, P. *Francis Bacon: From Magic to Science.* London: Routledge & Kegan Paul, 1968.
Russell, B. *The Problems of Philosophy.* Oxford: Oxford University Press, 1962.
Salmon, W. C. 'The Foundations of Scientific Inference' in R. Colodny (ed.), *Mind and Cosmos: Essays in Contemporary Science and Philosophy.* Pittsburgh: Pittsburgh University Press, 1966.
———. *Logic.* New Jersey: Prentice-Hall, 1973.
Sankey, H. 'In Defence of Untranslatability'. *Australasian Journal of Philosophy,* 68 (1990): 1–21.
———. 'Incommensurability and the Indeterminacy of Translation'. *Australasian Journal of Philosophy,* 69 (1991): 219–23.
———. 'Translation Failure Between Theories'. *Studies in History and Philosophy of Science,* 22 (1991): 223–36.
Savage, C. W. (ed.). *Scientific Theories. Minnesota Studies in the Philosophy of Science XIV.* Minneapolis: University of Minnesota Press, 1990.
Scheffler, I. *Science and Subjectivity.* Indianapolis: Bobbs-Merrill, 1967.
———. 'Vision and revolution: A Postscript on Kuhn'. *Philosophy of Science,* 39 (1972): 366–74.
Sciama, D. W. *Modern Cosmology.* Cambridge: Cambridge University Press, 1975.
Shapere, D. 'The Structure of Scientific Revolutions'. *Philosophical Review,* 73 (1964): 333–94. Reprinted in G. Gutting (ed.), *Paradigms and Revolutions: Applications and Appraisals of Thomas Kuhn's Philosophy of Science.* Notre Dame, Indiana: Notre Dame University Press, 1980.
———. 'Meaning and Scientific Change' in R. Colodny (ed.), *Mind and Cosmos: Essays in Contemporary Science and Philosophy.* Pittsburgh: Pittsburgh University Press, 1966. Reprinted in I. Hacking (ed.), *Scientific Revolutions.* Oxford: Oxford University Press, 1981.
Shapin, S. 'History of Science and its Sociological Reconstructions'. *History of Science,* 20 (1982): 157–211.

———. 'Following Scientists Around'. *Social Studies of Science,* 18 (1988): 533–50.

Siegel, H. 'Discovery, Justification and the Naturalizing of Epistemology'. *Philosophy of Science*, 47 (1980): 297–321.

———. 'Objectivity, Rationality, Incommensurabilty, and More'. *British Journal for the Philosophy of Science*, 31 (1980): 359–75.

———. 'Truth, Problem Solving and the Rationality of Science'. *Studies in History and Philosophy of Science*, 14 (1983): 89–112.

———. 'What is the Question Concerning the Rationality of Science?'. *Philosophy of Science*, 52 (1985): 517–37.

———. *Relativism Refuted: A Critique of Contemporary Epistemological Relativism.* Dordrecht: Reidel, 1987.

———. 'Farewell to Feyerabend'. *Inquiry*, 32 (1989): 343–69.

Simon, H. A. *et al. Scientific Discovery: Computational Exploration of the Creative Process.* Cambridge, Massachusetts: MIT Press, 1987.

Simon, H. A. 'Comments on the Symposium on "Computer Discovery and the Sociology of Scientific Knowledge"'. *Social Studies of Science*, 21 (1991): 143–8.

Skyrms, B. *Choice and Chance: An Introduction to Inductive Logic.* Belmont, California: Dickenson, 1986.

Slezak, P. 'Computers, Contents and Causes: Replies to My Respondents'. *Social Studies of Science*, 19 (1989): 671–95.

———. 'Scientific Discovery by Computer as Empirical Refutation of the Strong Programme'. *Social Studies of Science*, 19 (1989): 563–600.

Smart, J. J. C. *Between Science and Philosophy: An Introduction to the Philosophy of Science.* New York: Random House, 1968.

Smokler, H. 'Institutional Rationality: The Complex Norms of Science'. *Synthese*, 57 (1983): 129–38.

Stewart, J. A. *Drifting Continents and Colliding Paradigms: Perspectives on the Geoscience Revolution.* Bloomington and Indianapolis: Indiana University Press, 1990.

Stone, M. A. 'A Kuhnian Model of Falsifiability'. *British Journal for the Philosophy of Science*, 42 (1991): 177–85.

Storer, N. *The Social System of Science.* New York: Holt, Rinehart and Winston, 1966.

Stuewer, R. H. (ed.). *Historical and Philosophical Perspectives of Science. Minnesota Studies in the Philosophy of Science V.* Minneapolis: University of Minnesota Press, 1970.

Suppe, F. (ed.), *The Structure of Scientific Theories*. Urbana, Illinois: University of Illinois Press, 1977.

Suppes, P. 'Review of Larry Laudan, *Science and Values*'. *Philosophy of Science*, 53 (1986): 449–51.

Takeuchi, H., Uyeda, S. and Kanamori, H. *Debate About the Earth: Approach to Geophysics through Analysis of Continental Drift*. San Francisco: Freeman, Cooper & Co., 1970.

Taylor, C. 'Rationality' in M. Hollis, and S. Lukes (eds), *Rationality and Relativism*. Oxford: Blackwell, 1982.

Thagard, P. 'The Best Explanation: Criteria for Theory Choice'. *Journal of Philosophy*, 75 (1978): 76–92.

———. 'Against Evolutionary Epistemology' in P. D. Asquith and R. N. Giere (eds), *PSA 1980: Proceedings of the 1980 Biennial Meeting of the Philosophy of Science Association*, vol. 2. East Lansing, Michigan: Philosophy of Science Association, 1980.

———. *Computational Philosophy of Science*. Cambridge, Massachusetts: MIT Press, 1988.

———. 'Welcome to the Cognitive Revolution'. *Social Studies of Science*, 19 (1989): 653–7.

——— and Nowak, G. 'The Explanatory Coherence of Continental Drift' in A. Fine and J. Leplin (eds), *PSA 1988: Proceedings of the 1988 Biennial Meeting of the Philosophy of Science Association*, vol. 1. East Lansing, Michigan: Philosophy of Science Association, 1988.

Tianji, J. 'Scientific Rationality, Formal or Informal?'. *British Journal for the Philosophy of Science*, 36 (1985): 409–23.

Trigg, R. 'The Sociology of Knowledge'. *Philosophy of the Social Sciences*, 8 (1978): 289–98.

Turnbull, C. M. *The Forest People*. London: Jonathan Cape, 1961.

van der Vet, P. 'Overdetermined Problems and Anomalies'. *Studies in History and Philosophy of Science*, 10 (1979): 259–61.

van Fraassen, B. C. *The Scientific Image*. Oxford: Clarendon, 1980.

Wallis, R. (ed.). *On the Margins of Science: The Social Construction of Rejected Knowledge*. Keele: University of Keele, 1979.

Wegener, A. *The Origin of Continents and Oceans*, tr. from the 4th German edn by J. Biram. New York: Dover, 1966.

Westman, R. S. 'Towards a Richer Model of Man: A Critique of Laudan's *Progress and its Problems*' in P. D. Asquith and I. Hacking (eds), *PSA 1978: Proceedings of the 1978 Biennial Meeting of the Philosophy of Science Association*, vol. 2. East Lansing, Michigan: Philosophy of Science Association, 1981.

Whorff, B. *Language, Thought and Reality*, ed. J. Carroll. Cambridge, Massachusetts: MIT Press, 1956.

Wolfenstein, L. and Beier, E. W. 'Neutrino Oscillations and Solar Neutrinos'. *Physics Today*, July 1989.

Wood, R. M. *The Dark Side of the Earth*. London: Allen & Unwin, 1985.

Woolgar, S. 'Interests and Explanation in the Social Study of Science'. *Social Studies of Science*, 11 (1981): 365–94.

——— (ed.). *Knowledge and Reflexivity: New Frontiers in the Sociology of Knowledge*. London: Sage, 1988.

———. *Science: the very thing*. Chichester: Ellis-Horwood, 1988.

Worrall, J. 'Against Too Much Method. Review of P. K. Feyerabend *Against Method*'. *Erkenntis*, 13 (1978): 279–95.

———. 'The Value of a Fixed Methodology'. *British Journal for the Philosophy of Science*, 39 (1988): 263–75.

———. 'Fix it and be Damned: A Reply to Laudan'. *British Journal for the Philosophy of Science*, 40 (1989): 376–88.

———. 'Rationality, Sociology and the Symmetry Thesis'. *International Studies in the Philosophy of Science*, 4 (1990): 305–19.

Yearly, S. 'The Relationship between Epistemological and Sociological Cognitive Interests: Some Ambiguities Underlying the Case of Interest Theory in the Study of Scientific Knowledge'. *Studies in History and Philosophy of Science*, 13 (1982): 353–88.

Index